T0234270

Springer Series in Advanced Manufacturing

Series Editor

Duc Truong Pham, University of Birmingham, Birmingham, UK

The **Springer Series in Advanced Manufacturing** includes advanced textbooks, research monographs, edited works and conference proceedings covering all major subjects in the field of advanced manufacturing.
The following is a non-exclusive list of subjects relevant to the series:

1. Manufacturing processes and operations (material processing; assembly; test and inspection; packaging and shipping).
2. Manufacturing product and process design (product design; product data management; product development; manufacturing system planning).
3. Enterprise management (product life cycle management; production planning and control; quality management).

Emphasis will be placed on novel material of topical interest (for example, books on nanomanufacturing) as well as new treatments of more traditional areas.

As advanced manufacturing usually involves extensive use of information and communication technology (ICT), books dealing with advanced ICT tools for advanced manufacturing are also of interest to the Series.

Springer and Professor Pham welcome book ideas from authors. Potential authors who wish to submit a book proposal should contact Anthony Doyle, Executive Editor, Springer, e-mail: anthony.doyle@springer.com.

Myo Min Aung · Yoon Seok Chang

Cold Chain Management

 Springer

Myo Min Aung
Department of Mechatronics and Robotics
Engineering
Faculty of Technical Education
Rajamangala University of Technology
Thanyaburi (RMUTT)
Pathum Thani, Thailand

Yoon Seok Chang
School of Air Transport and Logistics
Korea Aerospace University (KAU)
Goyang, Korea (Republic of)

ISSN 1860-5168   ISSN 2196-1735 (electronic)
Springer Series in Advanced Manufacturing
ISBN 978-3-031-10376-6   ISBN 978-3-031-09567-2 (eBook)
https://doi.org/10.1007/978-3-031-09567-2

This Springer imprint is published by the registered company Springer Nature Switzerland AG
The registered company address is: Gewerbestrasse 11, 6330 Cham, Switzerland

Trust in the Lord with all your heart and lean not on your own understanding; in all your ways acknowledge him, and he will make your path straight. (Proverbs 3:5–6)

Whatever you do, do with all your heart. Wisdom makes light the darkness of ignorance. If you light a lamp for somebody, it will also brighten your path. Make of yourself a light. (Buddha)

Preface

In the era of globalization, the trade is rapidly growing not only in domestic level but also in international level across countries and continents. As long as the distance from the producer to the consumer increases, keeping quality, safety, and security of the goods along the chain has become a significant challenge. Perishable goods which are time and temperature sensitive in nature are of higher value and more vulnerable to temperature disturbances. The logistics industry needs more attention to handle perishables as they need quick and smart decision than non-perishables. For perishable goods, the cold chain remains one of the most important ways to preserve and deliver them to market in safe and good condition.

The shelf life, quality, and safety of perishable foods throughout the supply chain are significantly impacted by environmental factors, especially temperature. Not only wastage and loss of value but also quality, safety, and security are big concerns faced by food industry today. Cold chain is the solution to maintain the quality, the shelf life, and the safety of foods delivered to final consumers. Temperature abuse in the cold chain can cause microbial growth and spoilage of the food products, thus becoming the factor for causing foodborne illness. In pharmaceutical industry, improper storage and use of drugs, vaccines, and blood can be harmful to the patients. From security perspective, the threat of bioterrorism enforces all actors to establish and maintain records to enhance the security of the cold chain.

The recent and ongoing COVID-19 pandemic highlights the importance of cold chain in the fight for the infectious diseases and can be seen as a strong driver for successful vaccination that saves millions of lives around the world. Supply chain disruptions occurred around the world, and managing the supply chain (integration, efficiency, and agility, etc.) to perform well has become crucial in crisis situations. Managing the cold chain not only keeps products at the proper temperatures under right conditions but also increases supply chain efficiency and reduces costs. Cold chain management (CCM) deals with efficient monitoring, control, and organization of production and logistics regarding temperature. It comprises of planning and implementation of single processes and process steps as well as implementation of instruments and methods of process monitoring and control. The continuous quality monitoring and assessment, corrective and preventive procedure, and prompt actions are important to maintain the cold chain.

The emphasis is on food industry which is becoming more customer-oriented and need faster response to deal with food scandals and incidents. Good traceability systems help to minimize the production and distribution of unsafe or poor-quality products, thereby minimizing the potential for bad publicity, liability, and recalls. The current food labeling system cannot guarantee that the food is authentic, quality, and safe. Therefore, traceability is applied as a tool for assurance of safety and quality in order to gain consumer confidence. Traceability should be achieved not only for the physical movement of the products but also for quality deterioration at critical points along the chain.

It is found that there are not many books about cold chain/cold chain management available today. This book is designed to provide a concise understanding on cold chain management to undergraduate university students, business owners, supply chain managers, and operators in the logistics and supply chain sector of business. We believe that the readers will receive valuable insight into cold chain management, which includes topics such as concepts and practices, tools and techniques, equipment and facilities, temperature management, quality assessment, and traceability, among others.

Pathum Thani, Thailand   Myo Min Aung
Goyang, Korea (Republic of)   Yoon Seok Chang

Acknowledgements

Our dedication of writing this book is to compile scattered knowledge in literature about the principle and practice of cold chain, especially in management and operation perspectives. The invaluable experiences are gained through the tasks such as reviewing the literature, collecting information from cold chain operators, revising the chapters, and dealing with copyrights permissions, etc. Efforts were made to include useful facts, updated trends, and important topics. We hope this book can fulfill a gap of having only few resource books in cold chain literature expecting to be a useful guidebook for those who are interested to acquire knowledge on cold chain/cold chain management.

First, the authors would like to acknowledge our families, teachers, and friends and colleagues who are important in the first place; without their support, it will be difficult to see our continuous effort that turns into the fruit as a book.

Second, there are many other people who helped bring this book to fruition, and we are grateful to all of them. Especially, thanks are due to people, companies, and organizations who allowed to use their figures and tables in our book. Our UTAC researchers, G. H. Bong, DongWoo Son, and Namuook Kim, deserve a special mention for their efforts in smart refrigerator project. Although not mentioned here, there are other grateful people who help, support, and encourage in every step of the preparation process. Without their support, this work could not become to a completion and a reality.

Third, the editors of Springer Nature are very supportive throughout the preparation process to accomplish this book. We are very thankful to Springer Nature editorial team for their patience, thoughtful suggestions, help, and encouragement.

Fourth, this book is a compilation of the essential topics of cold chain with some research outputs. This research was a part of the project (ID: 1525011943) titled "Development of low-cost seafood storage showcase" funded by the Ministry of Oceans and Fisheries, Republic of Korea. We would like to acknowledge their support.

Contents

About the Authors

Myo Min Aung is a lecturer of Mechatronics and Robotics Engineering Department at the Faculty of Technical Education, Rajamangala University of Technology Thanyaburi (RMUTT), in Thailand. Formerly, he worked as a research professor at Korea Aerospace University (KAU) in South Korea after completing his Ph.D. in Logistics there. He is a certified Kaizen consultant of Myanmar Productivity Centre (MPC) and vice president of Yangon Region Computer Professionals Association (MCPA-Yangon) in Myanmar. His research interests include RFID and wireless sensor network (WSN), logistics, SCM, POM, automation, robotics, and IoT. e-mail: myomin_a@rmutt.ac.th

Yoon Seok Chang is a professor of School of Air Transport and Logistics at Korea Aerospace University. He is also a director of Ubiquitous Technology Application Research Center (UTAC) at Korea Aerospace University. He was a senior research associate of Auto-ID Center, Cambridge University, UK, and a senior application engineer of i2 technologies (currently Blue Yonder), US. He finished his Ph.D. from Imperial College London, UK, 1997. From 2010 to 2011, he was a visiting faculty of the department of computer science, Caltech, US, and from 2011 to 2013, he was a dean of information service at Korea Aerospace University, respectively. His research interests include RFID and sensor network, air cargo management, SCM/SCE, and product lifecycle management. e-mail: yoonchang@kau.ac.kr

Abbreviations

AAR	Accelerated aging rate
AIDC	Automatic identification and data capture
AOTT	Absolute Optimal Target Temperature
AS/RS	Automated storage and retrieval systems
BCG	Bacillus Calmette–Guérin
BIFTS	Blockchain–IoT-based food traceability system
BRC	British Retail Consortium
BSE	Bovine spongiform encephalopathy
CA	Controlled atmosphere
CAC	Codex Alimentarius Commission
CAGR	Compound Annual Growth Rate
CCL	Cold Chain Logistics
CCM	Cold chain management
CFC	Chlorofluorocarbon
CG	Centre of gravity
COOL	Country-of-Origin Labeling
DNA	Deoxyribonucleic acid
DT	Diphtheria and Tetanus toxoid
DTP	Diphtheria, Tetanus and Pertussis
EPC	Electronic product code
FDA	Food and Drug Administration
FEFO	First Expired First Out
FG	Freshness Gauge
FIFO	First In First Out
FMD	Foot and Mouth Disease
FSA	Food Standard Agency
FSMS	Food Safety Management System
FTD	Flexible Tag Data logger
GDP	Good Distribution Practice
GHG	Greenhouse gases
GHP	Good Hygiene Practices

GIS	Geographic Information System
GMOs	Genetically Modified Organisms
GMP	Good Manufacturing Practices
GPS	Global Positioning System
GSFI	Global Food Safety Initiative
GSP	Good Storage Practice
GTD	Global Trade Digitization
HACCP	Hazard Analysis Critical Control Point
Hib	*Haemophilus influenzae* Type b
IARW	International Association of Refrigerated Warehouses
IC	Intelligent container
IFS	International Food Standard
IIR	International Institute of Refrigeration
IoT	Internet of Things
IQF	Individual Quick Freezing
ISO	International Standard Organisation
LELO	Last Expired Last Out
LILO	Last In Last Out
LNG	Liquified Natural Gas
LQFO	Low Quality First Out
LSLO	Long Shelf-life Last Out
MAP	Modified Atmosphere Packaging
MMR	Measles, Mumps and Rubella
mRNA	Messenger RNA
OPV	Oral Polio Vaccine
OSHA	Occupational Safety and Health Administration
PCM	Phase change material
PCR	Perishable Cargo Regulations
PDA	Personal Digital Assistant
PDCA	Plan-Do-Check-Act
PDO	Protected Designation of Origin
PGI	Protected Geographical Indication
PV	Photovoltaic
QMS	Quality Management Systems
RFID	Radio Frequency Identification
RH	Relative humidity
ROI	Return on Investment
ROTT	Relative Optimal Target Temperature
RS	Remote Sensing
SCM	Supply chain management
SOP	Standard operating procedure
SQF	Safe Quality Food
SSFO	Short Shelf-life First Out
Td	Tetanus and Diphtheria toxoid
TI	Temperature Indicator

TQM	Total Quality Management
TRU	Transport Refrigeration Unit
TT	Tetanus toxoid
TTI	Time-Temperature Indicator
UHF	Ultrahigh Frequency
UNICEF	United Nations Children's Fund
USDA	US Department of Agriculture
VVMs	Vaccine Viral Monitors
WHO	World Health Organisation
WRAP	Waste Resources and Action Programme
WSN	Wireless sensor network
ZECC	Zero Energy Cool Chambers

List of Figures

List of Tables

Chapter 1
Introduction

1.1 Synopsis

The cold chain is a concept of managing the storage, transportation, and distribution of temperature-sensitive products. Cold chains are common in the food and pharmaceutical industries and also in some chemical shipments. The primary task of cold chain is to control and maintain the proper temperature in order to prevent the growth of microorganisms and deterioration of biological products during harvest/production, processing, storage, and distribution. The cold chain includes all segments in the supply chain from the producer to the consumer.

Therefore, it can be seen as a single entity since a breakdown in temperature control at any stage can impact the final quality and shelf life of the product. Thus, when a link of this "cold chain" fails, it inevitably results in a loss of quality and revenue and, in many cases, leads to spoilage. The final stage of cold chain, in food industry (i.e., the stage of consuming food), is central to any society and has a wide range of social, economic, and, in many cases, environmental consequences.

The management of food supply chain networks is complicated by specific product and process characteristics. Most food products are perishable, and their shelf life and quality can be significantly affected by temperature conditions in the supply chain. Most perishable products have a strict environmental requirement when transported by reefer container or truck or stored at the refrigerated warehouse. In fact, there is frequently a loss of value when goods are being transported.

One of the challenging tasks in today's food industry is controlling the product quality throughout the food supply chain. Typically, the perishable products move through often-complex distribution networks and dramatically changing environmental conditions. Quality degradation of perishable products can happen at any time along the chain from producer to consumer. Temperature is found as the most important factor for quality and prolonging or maintaining the shelf life of the perishables.

Optimizing temperature control in refrigerated truck, warehouse, and cold store is crucial in today's Cold Chain Logistics operation. Temperature control is very

© Springer Nature Switzerland AG 2023
M. M. Aung and Y. S. Chang, *Cold Chain Management*, Springer Series in Advanced
Manufacturing, https://doi.org/10.1007/978-3-031-09567-2_1

complex when mixed loading of perishables is frequently required. The biggest challenge is the diverse characteristics of perishable foods that demand different temperature requirements.

The other issue is how to evaluate or assess the actual quality of perishable commodities in cold chain. Quality needs to be assured that the products are kept at the prime conditions along the supply chain. The deterioration of perishable foods can lead to a decrease in the aesthetic appeal as well as a reduction in nutritional value. Sometimes, the degradation of foods is readily visible like changes of texture or discoloration, but there are some situations where the degradation might not be so readily visible. Therefore, the food industry needs appropriate tools and methods to overcome these challenges.

Moreover, the ability to know and assess quality changes in real time is essential and it allows to achieve a competitive advantage for cold chain operators and to gain consumer confidence as well. The supply chain management for temperature-sensitive goods requires fast decisions; goods are forwarded within hours or during limited life cycle.

In the development of a cold chain, the temperature control, the knowledge on product characteristics, the control procedures and practices, the tools and techniques, technical equipment and cold facilities, laws and regulations, and supply chain management are essentially important for cold chain operations. In addition, cold chain assessment, audits, and traceability are necessary management tools for unbroken cold chain.

Nowadays, wireless intelligent solutions can monitor the cold chain and reduce risks in transit—by land, sea, and air. The use of time–temperature indicators (TTIs) and the adoption of radio frequency identification (RFID) and wireless sensor network (WSN) in cold chain provide an efficient way to monitor and control this serious situation regarding the loss of perishable products during transportation and storage. The new technologies such as Internet of Things (IoT) and blockchain offer great potential to cold chain to operate in global scale and support to run the process in smooth and seamless manner as a result of combining science, people, process, and technologies.

Cold chain management (CCM) plays an important role in tackling the challenges to obtain the optimization of product quality, product safety, and minimization of wastage. CCM mainly focuses on temperature monitoring and control at each step within the production, storage and transportation chain on inner- and inter-operation levels. CCM system requires smart decision through monitoring and traceability to provide high level of service while keeping quality of products. Good CCM can offer the value-added products to the customer in high service quality and good profit by reducing the waste and operation cost.

1.2 Organization of the Book

There are nine chapters in this book. The first chapter gives an overview about the cold chain and the book's organization. Chapter 2 introduces the fundamentals of cold chain and cold chain management that cover definitions, features, structure, principles, regulations, and standards. Chapter 3 presents the development of cold chain in history, the impacts, and the research trends of cold chain. Chapter 4 explains essential aspects that are important to consider in cold chain management. Chapter 5 lists the cold chain monitoring tools that play an important role in maintaining a sustainable cold chain. Chapter 6 attempts to give the idea on how to manage temperature in cold chain with extension on methods to define optimal target temperature.

Chapter 7 mentions how quality is identified and assessment can be done by using appropriate methods. Chapter 8 discusses the important role of traceability in managing food safety and quality.

Chapter 9 presents the design and development of a smart refrigerator and is followed by a selection of questions (and answers) for each chapter for readers to check their understanding of the topics in the book. The book is completed by an Index.

Chapter 2
Fundamentals of Cold Chain Management

2.1 Cold Chain

Cold chain is not a temperature-controlled trailer, not a temperature-controlled facility, and not a temperature sensor application. It is more than warehouse temperature monitoring and can be defined as a temperature-controlled supply chain which relates to an uninterrupted or interdependent series of production, storage, transportation, and distribution activities that maintain perishable goods in a safe, wholesome, and good quality state from the primary production to the final consumption stage (Beasley 2002; Kacimi et al. 2009).

Cold chain can be seen as a science, a technology, and a process as it relates to understanding of the chemical and biological processes linked with perishability, a technology that rely on physical means to ensure appropriate temperature conditions along the supply chain, and a process that consist of a series of tasks to prepare, store, transport, and monitor temperature-sensitive products (Rodrigue 2020). Cold chains are common in the food and pharmaceutical industries but also found its application in semiconductors, biologics, and some chemical shipments.

It is also named as low-temperature logistics system and is comprised of equipment and processes which keep perishable products under conditioned environment. Perishable products can be categorized into two types: living products (fruits, vegetables, live seafood, flowers, etc.) and non-living products (meat, dairy products, processed food products, medicines, blood, frozen products, etc.), which all require appropriate atmosphere to defy microbial spoilage (Donselaar 2006). The cold chain includes all aspects of the transfer of goods from the producer to the consumer. Logistics is the key of the cold chain. Therefore, many scholars call it "cold chain logistics". It includes refrigeration, freezing, warehousing, transportation, distribution, packaging, processing, multimodal transport, and other series of value-added services (Guojun and Rong 2009; Ma and Guan 2009).

© Springer Nature Switzerland AG 2023
M. M. Aung and Y. S. Chang, *Cold Chain Management*, Springer Series in Advanced Manufacturing, https://doi.org/10.1007/978-3-031-09567-2_2

Successful cold chains require effective planning, quick communication, the proper facilities, access to real-time information, and the right technology at every step of the journey. The logistical planning, refrigerated storage and packaging technologies, temperature management solutions that offer visibility and effective communication are essential to protect the integrity of cold chain.

Cold chain is essential for the supply chain management of the food and pharmaceutical industries because a good management of cold chain can reduce the risk and cost. The loss of a trailer of temperature-sensitive foods due to improper handling or transport will cost thousands of dollars: a pharmaceutical shipment, in millions (Kevan 2005). Perishable products must be continuously monitored for safety concerns throughout the whole supply chain. A breakdown in temperature control at any stage will impact on the final quality of the product (SARDI 2006).

From the factory to the consumers of food, products follow complex logistic circuits that are subjected to intrinsic constraints. First, the required chilling time between harvest, or the end of cooking, and loading is a constraint encountered by the producer and the carrier. The carrier's liability comes into play from the moment the products are taken over, and it is up to him to check the temperature manually during loading products, then reach the temperature-controlled distribution hub and send to the customer via supermarket or store. Temperature control needs to be improved throughout the cold chain to ensure food safety and hygiene and to maintain the product quality (Commere 2003). The sequence of events within a typical cold chain is the same as typical supply chain except essential consideration on required environmental condition (e.g., temperature, humidity, etc.) as illustrated in Fig. 2.1.

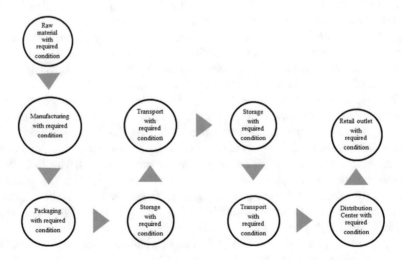

Fig. 2.1 Typical cold chain

2.2 Cold Chain Management

Cold chain management (CCM) refers to maintaining the proper temperature of the products through all the handoffs in the cold chain until it reaches the consumer (Smith 2005). The aim is to preserve quality of perishables and deliver them to market in safe and good condition. Therefore, the key to managing the cold chain is to monitor and maintain the product temperature and coordination in each and every stage of the supply chain. Figure 2.2 shows the concept of CCM.

In a cold chain, the shelf life, quality, and safety of perishable foods throughout the supply chain are significantly impacted by environmental factors, especially temperature. If the temperature of some chilly foods exceeds specific limits, the rise in temperature of just a few degrees can cause microbial growth leading to the great decrease of quality, spoilage of foods, and the increase of the risk of food poisoning (Carullo et al. 2008). Loss and damage of perishable goods during storage and transportation is a substantial global issue (Ruiz-Garcia and Lunadei 2010).

Temperature control in food supply chain is the most important factor to prolong the practical shelf life of produce. The first appearance of unsightly yellowing of broccoli, for example, may be delayed by three or more days if they are kept under refrigeration. Maintaining the desired or ideal holding temperature is a major factor in protecting perishable foods against quality loss and wastage. Temperature control in cold chain preserves both sensory and nutritional qualities; for example, vitamin C losses in vegetables can be up to 10% per day when stored at a temperature of 2 °C; however, vitamin C loss can increase to over 50% per day when stored at temperatures of +20 °C. Sweet corn may lose half of its initial sugar content in one

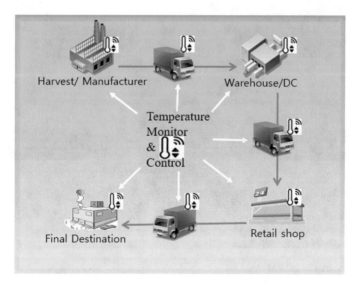

Fig. 2.2 Typical cold chain management

day at 21 °C, but only 5% of it at 0 °C. Most of the mechanisms of quality loss are determined by storage temperature and are accelerated with time spent above the recommended value. They are also promoted by temperature fluctuations (Stoecker 1998; Martin and Ronan 2000).

2.3 Features of Cold Chain Logistics

Cold Chain Logistics is complicated and challenging to maintain unbroken cold chain. As described by Dingyi (2010), the timing, the quality of products, the temperature, the humidity, and the environment are all important to Cold Chain Logistics. The three main features are:

Complexity

The cold chain is quite complex due to limited lifetime and the deteriorating quality of goods over time unless proper temperature level is not maintained (Bowman et al. 2009). There are challenges on how to deliver goods with long quality shelf life. Changes in original time (shelf life) due to temperature abuse will lead to decrease in quality that it is irreversible. And the different products have the different shelf life and temperature requirements that add substantial complexity to control. Temperature control in refrigerator or other cold storage facilities is also very complex to manage due to their different requirements and handling procedures. Figure 2.3 shows the factors that make the complexity of cold chain.

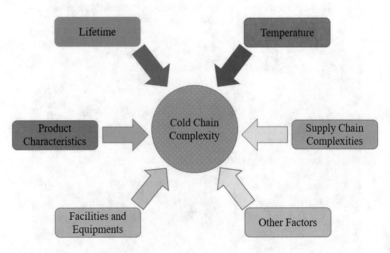

Fig. 2.3 Complexity of cold chain

Coordination

Perishable products are difficult to store unlike non-perishable one due to their biological nature; therefore, in every aspect of the logistics, process must be coordinated. The handling of goods can change hands many times along the chain. Everyone who deals with the cold chain has the responsibility to coordinate for unbroken and sustainable cold chain. Good coordination among supply chain nodes/parties in both information and physical distribution is essential for the successful CCM. For example, when a product is picked from a manufacturing plant to load to the delivering truck, careful coordination should be given in order to keep the product quality during transition. Also, when the product is delivered to the cold storage warehouse, the goods should be put into storage immediately; otherwise, it may lead to decline in quality. The important point is to follow standard practice and guidelines with good coordination among supply chain parties for the stable operation of the Cold Chain Logistics.

High Cost

Cold Chain Logistics has high investment in refrigeration equipment, telecommunication equipment, telecommunication fee, and insulation technologies that are obligatory. The cost with the warehouse and vehicles of Cold Chain Logistics is 3–5 times of the general dry products warehouse and vehicles. The investment in question involves, at its minimum, a main storage hub and a cold storage transportation vehicle. Cold Chain Logistics transportation costs are also high because electricity and oil fees are important prerequisite investment in the Cold Chain Logistics. The annual operating costs for cold chain businesses are much higher per cubic feet. Energy expenses alone account for 30% of the total expenses.

2.4   Structure of Cold Chain Logistics

Changsheng (2007) summarized that there are four parts of Cold Chain Logistics: cold processing, cold storage, and cold transportation and distribution.

Cold Processing

It includes cooling and freezing of meat and aquatic products, pre-cooling of fruits and vegetables, and low temperature processing of dairy products. In this part of Cold Chain Logistics, the necessary equipment is the refrigerating machineries and the quick-frozen machineries where the former lowers the products to their preferred temperature and then the latter would freeze them for longtime storage.

Cold Storage

It includes the freezing and storing of foods and controlled atmospheric storage of fruits and vegetables. It can also ensure the low temperature processing environment for food and other perishable products. In this part of the Cold Chain Logistics,

Fig. 2.4 Cold chain container (Reprinted by permission of Envirotainer—https://www.envirotai ner.com/)

the necessary equipment involved are the refrigerator, the freezer and the domestic refrigerator, and the other equipment. As mentioned by Kader (2004), controlled atmospheric storage is for the protection of fresh fruits, vegetables, and their products throughout postharvest handling. In this process, smart design and operation is required for saving electricity. Each product has different requirements for its storage temperature and operation. Many warehouses suffer from profitability due to both the poor design of cold storage and poor management of storage causing high electricity bills.

Cold Transportation and Distribution

There are different methods of cold transportations: refrigerated railway carriages, refrigerator vehicles, refrigerated ships, refrigerated containers, and other low temperature transports. Figure 2.4 shows an example of cold chain container by Envirotainer. As described by Foreinio and Wright (2005), supply chain efficiency can be increased by managing the Cold Chain Logistics, thus lowering its costs for the firm operating cold chain.

The principal functions of packaging in logistics operations are to include containment, protection, communication, and convenience. The logistical planning and thermal and refrigerated packaging methods are very important since cold chain refers to the transportation of temperature-sensitive goods from manufacturer to the consumer through appropriate distribution channels. The technological developments in packaging can offer the benefits of reducing losses, maintaining quality, adding value, and extending shelf life of agricultural produce and consequently secure the food system (Opara and Mditshwa 2013). There are two types of cold chain packaging systems: active and passive, which have different characteristics and are simplified by Roskoss (2011), Romero (2013) and Klinge Corp (2020).

Active thermal systems do not use any phase change materials (PCMs) such as water/ice or dry ice. These systems use mechanical- or electric-powered systems powered by an energy source, combined with thermostatic control to maintain proper product temperature. Active systems consist of cold chain transport containers featuring advanced electric- or battery-powered temperature controls. A system pushes cool air from the refrigeration system into the main payload area. Active systems work better for larger shipments and offer greater security, which minimizes the risk of theft. Active refrigeration systems, in addition, offer greater security as they are not limited like dry ice systems to operation only for a certain amount of time until the dry ice would all be consumed and need to be replaced.

Active systems are commonly found using in full trailer or reefer systems and provide heating, cooling, or both. Active systems often require limited preparation and are often simple to pack following a standard operating procedure (SOP). They are, however, dependent on the logistics of being brought to the site, packed, and collected within a fixed timeframe. The systems are high cost, and unless there is a full load, the efficiency of use may be restricted.

In contrast, passive thermal systems commonly use PCMs such as water/ice or dry ice. These shipping systems are the most basic and cost-effective. Passive packaging systems typically include polyurethane or polystyrene insulation or vacuum-insulated panels that can keep the product at a predetermined temperature for up to 96 h or longer. Passive systems can maintain tighter temperatures than active systems and are not susceptible to internal freezing. Passive systems can utilize technologies across a range of budgets and performance capabilities. They can provide frozen protection below -20 °C by using either the dry ice sublimation phase change at -78 °C, or the melting phase change of tailored materials to keep temperatures above -40 °C. There are also passive systems that can maintain $+2$ to $+25$ °C and $+15$ to $+25$ °C, using both water and specialized PCMs.

Disadvantages of passive systems include the need to condition the refrigerant to specific requirements and more complex shipping configurations as well as the primary constraint of time. Passive systems will fail at a 100% rate to maintain temperature if they do not arrive at the destination within the required packaging specification. In addition, passive solutions do not actively respond to adverse exposures and are specified to passively handle extreme exposures.

In addition, it may have another type for specific need in which two are combined as a hybrid system. Hybrid thermal systems use a combination of PCM such as water/ice or dry ice and thermostatic controls. Active systems often have built-in cooling units or rely on dry ice as a coolant in hybrid pattern.

When making the choice between active and passive systems, an appreciation of the product and chosen distribution model will have an important impact. The value of the product and the cost of replacement will also influence the decision, along with the routes, necessary thermal protection, and product volumes. The comparison of the cold chain packaging systems is shown in Table 2.1.

2.5 Principles of Cold Chain Logistics

As summarized by Jian (2010), five principles dominate the control of quality and safety of goods in Cold Chain Logistics. They are '3P', '3C', '3T', '3Q', and '3M' principles.

Table 2.1 Comparison of cold chain packaging systems

Active thermal system	Passive thermal system	Hybrid thermal system
Highest cost	Lowest cost	Moderate to high cost
Highest temperature accuracy	Lowest temperature accuracy	Moderate to high temperature accuracy
High reliability	High reliability with proper procedures	Moderate reliability with proper procedures
Standard/off the shelf sizes only—no custom sizes feasible	Standard/off the shelf and custom sizes- may not require tooling	Standard/off the shelf sizes only—no custom sizes feasible
Easy to use	Phase change material conditioning required—hardest to use	Phase change material conditioning required—added complexity

'3P' Principle

'3P' is extended as **Produce, Processing,** and **Package**. It requires the good quality of raw material produce, the high technology processing, and the packaging must be suitable to the products' properties because different products should be used with different packages. For example: Meats widely used vacuum package. Ice cream widely used paper box, plastic bucket, etc. This is the quality of the products' early management in the Cold Chain Logistics.

'3C' Principle

Employees should **Care** for the products, keep a **Clean** environment and keep the environment **Cool** in the whole process. These are fundamental conditions to guarantee the quality of products.

'3T' Principle

The quality of products depends on low temperature storage and the **Time, Temperature,** and **Tolerance** (TTT) of the transportation. There is a relationship between the time maintained in cold storage and the product's temperature. TTT factors maintain quality and safety during storage and offer guidance on how to deliver foods with long quality shelf life. TTT concepts refer to the relationship between storage temperature and storage life. For different foods, different mechanisms govern the rate of quality degradation and the most successful way of determining practical storage life is to subject the food to long-term storage at different temperatures. TTT relationships also predict the effects of changing or fluctuating temperatures on quality shelf life. As a guide to food manufacturers, the International Institute of Refrigeration has published "Recommendations for the processing and handling of frozen foods (2006)" (commonly known as the "Red Book"), which gives indications of recommended storage life for different foods (Martin and Ronan 2000).

'3Q' Principle

This principle is defined as the **Quantity** and **Quality** of equipment, and the **Quick** operation organization in the Cold Chain Logistics. The right quantity and good quality of the equipment and quickly operated organization can consistently guarantee the products always in a suitable process environment. Quick operation organization depends on the production process, transport vehicle for the transportation, and storage capacity for storing the products, etc. All these tasks should be quickly coordinated.

'3M' Principle

It means that the **Means**, **Methods**, and **Management** of storage. In the Cold Chain Logistics, the company should use appropriate transport machinery and storage methods understanding the diverse characteristics of all kinds of products. It will make the management more efficient in the Cold Chain Logistics.

2.6 Supply Chain Versus Cold Chain Management

It is generally said that CCM is one of the branches of supply chain management. It has the elements like in a typical supply chain: the origin and destination, the product to move, and the distribution process but rely on cooling technology to maintain the integrity of shipped goods. The purpose of the CCM is managing activities related to perishable products like medicine, blood, dairy, meat, food, vegetables, mushrooms, flowers and fruit products, and so on which must be distributed in a special timeframe and kept in the product-specific conditioned environment (Wen and Ouyang 2013).

Kovacs (2008) states that supply chain is a "product-based approach" of industrial ecology. It includes the companies that provide the products, raw materials, technology, and service to make the operation of the production chain smooth. Maxwell et al. (2006) argues that both supply chain and CCM could be defined under the "umbrella" of the sustainable production and coordination.

Therefore, the two main differences between supply chain and the cold chain are: First, compared with supply chain, the cold chain demands a lot on the operating conditions; second, from the production spots to the consuming place, products in the cold chain have the possibility to spoil (Joshi et al. 2009). Meanwhile, the relationship between supply chain and cold chain is that cold chain can be viewed as facilities and conditions demanded in a supply chain (Salin et al. 2002), and sustainability is important for both (Maxwell et al. 2006).

2.7 Global Cold Chain Management

Due to the globalization, there are an increasing number of global food companies that import and export food products. More goods are being transported further and frequently than ever before, and distance regarding food supply chain has drastically increased in recent years (Wallgren 2006). Advances in transportation and refrigeration technology have made it possible for shippers to deliver perishable products to purchasers thousands of miles away with no substantial loss in freshness and quality and at lower and lower costs. Markets and Markets (2020), a global market research and consulting company forecasts that the cold chain market is estimated to account for revenue of USD 233.8 billion in 2020 and is projected to grow at a Compound Annual Growth Rate of 7.8% to reach a value of USD 340.3 billion by 2025.

Bogataj et al. (2005) defined global CCM as "the process of planning, implementing, and controlling efficient, effective flow and storage of perishable goods, related services and information from one or more points of origin to the points of production, distribution, and consumption in order to meet customers' requirements on a worldwide scale". It is the process of integrating the existing business activities, including special activities for perishable goods conservation along the value chains, where more suppliers of certain raw materials or more production cells of certain semi-products (e.g., ready meals) appear in order to create value for the end user.

The global cold chain market has also been witnessing the increasing demand for cold chain logistics in emerging economies. However, reliable infrastructure, regulatory requirements, and energy costs are seen as a challenge to the growth of this market. Also, there remain significant constraints to the expansion of perishable product trade. Some constraints derive from economic and environmental issues associated with the technologies. The availability of economical and environmentally friendly refrigerants for reefer systems and inexpensive controlled atmosphere (CA) technologies is important for carriers to adopt and install. Moreover, efficient inspection and customs services by government agencies, as well as port-to-market distribution systems, are critical (Coyle et al. 2001).

2.8 Cold Chain Regulations and Standards

Many process industries, notably those involving food and beverages, drugs, cosmetics, and medical apparatus, are subject to government regulation and must maintain records that detail the lot identification of materials used in the manufacture of these products. Control of the cold chain is vital to preserve the safety and quality of refrigerated foods and comply with legislative directives and industry "codes of practice" (Martin and Ronan 2000).

In Europe, the EU directive 178/2002 went into effect on January 1, 2005, and requires mandatory traceability for all food and feed products sold within European Union countries (Folinas et al. 2006). In US, Bioterrorism Act of 2002 went into effect on December 12, 2003, and requires persons who "manufacture, process, pack, transport, distribute, receive, hold, or import food" to establish and maintain records. It also allows Food and Drug Administration to inspect those records if there is a reasonable belief that an article of food presents a serious health threat (Levinson 2009).

The World Health Organization working document QAS/04.068 also stated that special storage conditions (e.g., temperature and relative humidity) are required during transit; these should be provided, checked, monitored, and recorded to ensure the quality and integrity of pharmaceutical products during all aspects of the distribution process. In order to maintain the original quality, every activity in the distribution of pharmaceutical products should be carried out according to the principals of Good Manufacturing Practice, Good Storage Practice, and Good Distribution Practice (Bishara 2006). Therefore, these legislative directives enforce the supply chain companies to implement traceability and to ensure the quality and safety of the product through the stages of production, processing, and distribution.

Specifically, the Perishable Cargo Regulations manual published by International Air Transport Association is the leading publication for temperature control and CCM of goods from the health care and food sectors, including pharmaceutical products and non-hazardous biological materials. The PCR includes information on the latest efficient practices and everything needed to properly prepare, package, and handle time- and temperature-sensitive goods quickly and efficiently (https://www.iata.org/).

References

Beasley SD (2002) Helping exports keep their cool. AgExporter 14(4):4–5

Bishara RH (2006) Cold chain management—an essential component of the global pharmaceutical supply chain. Am Pharm Rev. Available online via http://intelsius.com/wp-content/uploads/2011/10/Pharma-Cold-Chain-Bishara_APR.pdf. Accessed 20 Apr 2021

Bogataj M, Bogataj L, Vodopivec R (2005) Stability of perishable goods in cold logistics chains. Int J Prod Econ 93–94:345–356

Bowman AP, Ng J, Harrison M, Lopez TS, Illic A (2009) Sensor based condition monitoring. Bridge Project. www.bridge-project.eu

Carullo A, Corbellini S, Parvis M, Reyneri L, Vallan A (2008) A measuring system for the assurance of the cold-chain integrity. In: Proceedings of the IEEE international instrumentation and measurement technology conference, Vancouver, Canada, pp 1598–1602

Changsheng B (2007) On operation management of cold chain. Doctoral's thesis. Tongji University, China

Commere B (2003) Controlling the cold chain to ensure food hygiene and quality. Bull IIR—No. 2003-2

Coyle W, Hall W, Ballenger N (2001) Transportation technology and the rising share of US perishable food trade. In: Changing structure of global food consumption and trade/WRS-01-1. Economic Research Service/USDA, pp 31–40

Dingyi D (2010) China cold chain logistics development report. China Logistics Publishing House

Donselaar K, Woensel T, Broekmeulen R, Fransoo J (2006) Inventory control of perishables in supermarkets. Int J Product Econ 104(2):462–472

Folinas D, Manikas I, Manos B (2006) Traceability data management for food chains. Br Food J 108(8):622–633

Foreinio H, Wright C (2005) Cold chain concerns. Pharm Technol 4:44–50

Guojun J, Rong G (2009) Research on the security of cold-chain logistics. In: 6th international conference on service systems and service management (ICSSSM), pp 757–761

Jian Z (2010) The process study of the cold-chain logistics center layout and the planning of cold storage, Master's thesis. Southwest Jiaotong University, China

Joshi R, Banwet DK, Shankar R (2009) Indian cold chain: modeling the Inhibitors. Br Food J 111(11):1260–1283

Kacimi R, Dhaou R, Beylot AL (2009) Placide: ad hoc wireless sensor network for cold chain monitoring. In: Performance modeling and analysis of heterogeneous networks. River Publishers, Denmark, pp 153–167

Kader A (2004) Brief description of controlled atmosphere storage. University of California

Kevan T (2005) Control of the cold chain. Frontline Solutions

Klinge Corp (2020) What is the cold chain process? https://klingecorp.com/blog/what-is-the-cold-chain-process/. Accessed 20 Apr 2021

Kovacs G (2008) Corporate environmental responsibility in the supply chain. J Clean Prod 16:1571–1578

Levinson DR (2009) Traceability in the food supply chain. Retrieved from http://oig.hhs.gov/oei/reports/oei-02-06-00210.pdf. Accessed 12 Aug 2020

Ma G, Guan H (2009) The application research of cold-chain logistics delivery schedule based on JIT. In: International conference on industrial mechatronics and automation, pp 368–370

Markets and Markets (2020) Retrieved from http://www.marketsandmarkets.com/Market-Reports/cold-chains-frozen-food-market-811.html. Accessed 12 Apr 2020

Martin G, Ronan G (2000) Managing the cold chain for quality and safety. Flair-flow Europe technical manual 378A/00. https://seafood.oregonstate.edu/sites/agscid7/files/snic/managing-the-cold-chain-for-quality-and-safety.pdf. Accessed 12 Apr 2021

Maxwell D, Sheate W, Vorst RVD (2006) Functional and systems aspects of the sustainable product and service development approach for industry. J Clean Prod 14:399–416

Opara UL, Mditshwa A (2013) A review on the role of packaging in securing food system: adding value to food products and reducing losses and waste. Afr J Agric Res 8(22):2612–2630

Rodrigue J-P (2020) The geography of transport systems, Appendix B-B9, 5th edn. Routledge

Ruiz-Garcia L, Lunadei L (2010) Monitoring cold chain logistics by means of RFID. In: Turcu C (ed) Sustainable radio frequency identification solutions, Croatia, Intech, pp 37–50

Romero B (2013) Cold chain packaging systems: comparisons of active, passive and hybrid thermal systems. https://www.pharmalogisticsiq.com/logistics/articles/cold-chain-packaging-systems-comparison-of-active. Accessed 20 Apr 2021

Roskoss A (2011) Temperature-controlled packaging systems: active or passive? Innov Pharma Technol 37

SARDI (2006) Maintaining the cold chain: air freight of perishables. South Australian Research and Development Institute

Salin V, Nayga RM (2002) A cold chain network for food exports to developing countries. Int J Phys Distrib Logist Manag 33(10):918–933

Smith JN (2005) Specialized logistics for a longer perishable supply chain. World Trade 18(11):46–48

Stoecker WF (1998) Refrigeration and freezing of food. Industrial refrigeration handbook. McGraw Hill

Wallgren C (2006) Local or global food markets: a comparison of energy use for transport. Local Environ 11(2):233–251

Wen L, Ouyang M (2013) Risk evaluation of cold logistics chain based on cloud model. Res J Appl Sci Eng Technol 5(6):2019–2026

Chapter 3
The Development of Cold Chain

3.1 Historical and Modern Development

Prior to mechanical refrigeration systems, people harvested natural ice and used cellars to cool and preserve the perishables, especially food. There were other preservation methods available: salting, pickling, drying, spicing, and smoking. However, most of the fresh foods and other temperature-sensitive goods were available only in local market due to their short shelf life and perishability. With the advancement of cooling technology, temperature-controlled supply chain or cold chain is developed; therefore, the fresh food today can be stored longer and transported to different markets across the borders of the countries. Especially, refrigeration technologies play vital role to preserve and transport perishable food from the point of production to the point of consumption. In the past, a cold chain simply denoted single temperature warehouses and refrigerated vehicles. There was no awareness of integrating the supply chain links and as a result billions of dollars' worth of losses occurred every year (Beasley 2002).

The refrigerated movement of temperature-sensitive goods is a practice that dating back to 1797 when British fisherman used natural ice to preserve their fish stockpiles (Wang and Wang 2005). It was also seen in the late 1800s for the movement of food from rural farms to urban centers (Cleland 1996). Cold storage was also a key component of fresh food transmission between colonial powers and their colonies. For example, in the late 1870s and early 1880s, France was starting to receive large shipments of frozen meat and mutton carcasses from South America, while Great Britain imported frozen beef from Australia and pork and other meat from New Zealand. By 1910, 600,000 tons of frozen meat was being brought into Great Britain alone (James et al. 2006).

Through technological development, cold stores (in Europe) or refrigerated warehouses (in US) have been built for handling temperature-controlled perishables. In addition, multi-commodity cold stores with chambers for different temperature ranges and atmosphere requirements are equipped for professional handling for a

© Springer Nature Switzerland AG 2023
M. M. Aung and Y. S. Chang, *Cold Chain Management*, Springer Series in Advanced
Manufacturing, https://doi.org/10.1007/978-3-031-09567-2_3

wider range of product sets. The refrigeration system is designed to adjust and operate to a range of temperature and humidity conditions, depending on the compatibility group of products (NHB 2010). IIR/UNEP (2007) reported that there were about 1300 million household refrigerators (more than 80 million are produced annually), 340 million air-conditioning units, and 350 million m^3 of cold storage facilities operating worldwide.

Further development of the perishables trade has led to the advanced transportation system (James et al. 2006). It was estimated by Heap (2006) that there are approximately 1300 specialized refrigerated cargo ships, 80,000 refrigerated railcars, 650,000 refrigerated containers, and 1.2 million refrigerated trucks in use worldwide. Jedermann et al. (2007) presented that the refrigerated containers can be equipped with radio frequency identification (RFID), wireless sensor network (WSN) and agent systems to assist monitoring, tracking, and decision-making during transport. By integrating the intelligent container (IC) with the Internet of Things (IoT), the lack of information will be eliminated, and thus, wastes of food will be reduced (Dittmer et al. 2012).

Management of the cold chain has also improved. Designs for intelligent cooling and storage system are implemented, and separate cooling and storage functions for specific products are developed. Concerning "temperature disturbance" issues, distributors, and retailers came to a consensus of not checking goods until they had been transferred into the temperature-controlled chambers at the store, which improved cold chain integrity. Retail products were delivered in plastic trays, on "dollies", or on roll cages, which improved handling speed (Fernie and Sparks 2004).

Another development of the cold chain occurred between the 1980s and 2000s when the ordering and replenishment cycle was shortened and the ordering quantity largely shrunk (McKinnon and Campbell 1998). Now, it become a practice that there is no fresh stock held in the cold distribution center for more than one day and stock holding in frozen products has declined to less than 10 days (IGD 2001).

Refrigeration is the best technology to date, with no associated risks, to ensure the safety of foods through chilling or freezing. However, compliance needs to be ensured at all stages in the cold chain, without which foodborne diseases and series of deaths occur, even in the most developed countries (Coulomb 2008). Therefore, a concept of "cold traceability" was introduced to trace groups of temperature-sensitive products like meat, fish, fruit, and pharmaceuticals which are transported in different atmosphere requirements. Bogataj et al. (2005) presented that this is enabled by some specific tools such as thermometers, RFID, and time–temperature integrators (TTI).

Recent developments and advances show that the demand for sustainable cold chain is still growing; innovations and development in packaging, fruit and vegetable coatings, bio-engineering (controlled ripening), logistics automation, transport technologies, cold traceability, energy efficiency, renewable energy, and other emerging cold chain technologies will reduce the deterioration of food products and help shippers to extend the reach of perishable products (Opara and Mditshwa 2013; Mercier et al. 2017; Bharti 2017; Onwude et al. 2020).

3.2   Food Cold Chain

A typical food supply chain consists of stages starting from farming/original production. The second stage typically goes to processing and packaging that involves with the next stage, storage, and distribution. The distribution stage starts with the wholesaler which delivers the goods to the customer/consumer usually via a retail stage. The imports and exports functions usually involve in these stages providing necessary inputs/facilities or exporting goods to deliver to markets. A schematic diagram of a food supply chain is shown in Fig. 3.1.

A food supply chain is complex, time-critical, and dynamic (Bourlakis and Weighman 2001; Olsson 2004; Trienekens et al. 2012). The food industry is facing challenges due to increasing operational complexity, frequently changing consumer needs, government regulations, and short product life cycles. It requires a very smart, efficient, and agile supply chain to manage the ever-changing needs of the end customers.

A food supply chain or food system refers to the processes that describe how food from a farm ends up on our tables. The processes include production, processing, distribution, consumption, and disposal. Food supply chain is domino-like, when one part of the food supply chain is affected, the whole food supply chain is affected, which is often manifested through changes in price. Movements of food and money are facilitated by "pulls" and "pushes" (CHGE 2012). To maintain the cold chain, every stage is important and cold storage facilities play essential role from harvest time to dining time. After harvest/slaughtering, the products should be kept under controlled temperature conditions to keep freshness, a significant quality parameter. Figure 3.2 shows a model of solar-powered cold storage container that can be used in

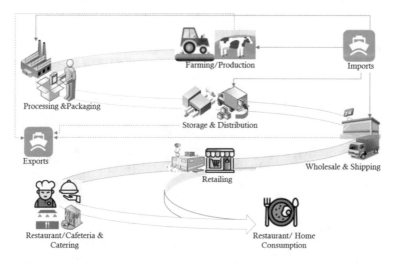

Fig. 3.1 Food supply chain schematic

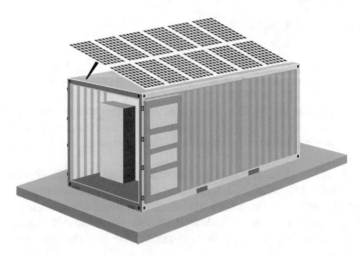

Fig. 3.2 Model of solar-powered cold storage container

agricultural farms for temporary storage of agricultural products after harvest before going to the processing stage.

Food can be classified into fresh or processed, perishables or non-perishables, and plant origin or animal origin. Perishable foods are foods that will spoil, rot, go bad very fast unless we keep them under refrigeration or freezing (e.g., fresh produce), while non-perishables are items that do not spoil or decay easily (e.g., dry and canned foods). The classification of foods into various groups can be seen in Fig. 3.3.

Most food products are perishable, and their shelf life can be greatly affected by temperature conditions in the supply chain. Time/temperature control becomes a critical issue in fresh food logistics, and the efficient and effective tracking of cold chain conditions is one of the main points to be addressed (Montanari 2008). Inadequate temperature is one of the factors causing foodborne illness. Temperature control is the most important factor to prolong the practical shelf life of produce and

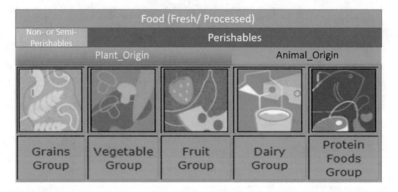

Fig. 3.3 Food groups

to reduce foodborne illness. Maintaining the desired or ideal holding temperature is a major factor in protecting perishable foods against quality loss. The complexity of food supply chain increases further because of safety and regulatory requirements, short shelf life of products, product recalls, traceability requirements, and quality assurance.

Quality deterioration can happen in foods even under the controlled temperature conditions: refrigeration and freezing. The frozen foods that deteriorate during storage by different modes or mechanisms are summarized in Table 3.1. Microbes usually are not a problem since they cannot grow at freezing temperatures. Enzymes are a big concern for frozen foods, which can cause flavor change (lipoxygenase) in non-blanched fruits and vegetables and accelerated deterioration reactions in meat and poultry (enzymes released from disrupted membranes during precooking). Cell damage or protein and starch interactions during freezing cause drip and mushiness upon thawing. Discoloration could occur by non-enzymatic browning, bleaching, and freezer burn. Vitamin C loss is often a major concern for frozen vegetables. Physical changes, such as package ice formation, moisture loss, emulsion destabilization, recrystallization of sugars, and ice of frozen desserts are often accelerated by fluctuating temperatures (Fu and Labuza 1997).

Table 3.1 Deterioration modes of frozen foods	Deterioration modes	Categories of frozen foods
	Rancidity	Meat, poultry
		Seafood and fish
		Convenience foods
	Toughening	Meat, poultry
		Seafood and fish
	Discoloration	Meat, poultry
		Seafood and fish
		Fruits and vegetables
		Concentrated juices
		Convenience foods
	Loss of flavor (lipoxygenase, peroxidase)	Fruits and vegetables
		Dairy products (ice cream and yogurt)
		Concentrated juices
		Convenience foods
	Loss of texture	Seafood and fish
		Fruits and vegetables
	Nutritional loss (vitamins and minerals)	Seafood and fish
		Fruits and vegetables
		Concentrated juices

Food safety and food quality are two important terms which describe aspects of food products and the reputations of the processors who produce food. CAC (2003) defines food safety as an assurance that food will not cause harm to the consumer when it is prepared and/or eaten according to its intended use. Food safety refers to all hazards, whether chronic or acute, that may make food injurious to the health of the consumer. It is not negotiable and a global issue affecting billions of people who suffer from diseases caused by contaminated food. Both developed and developing countries share concerns over food safety as international food trade and cross-border movements of people and live animals increase (Asian Productivity Organisation 2009). In industries such as telecommunications, software development and airlines, security is the principal driver for traceability in contrast to food industry where the safety is a really important issue (Opara 2003). The traceability of food products and the ability of food facilities to provide information about their sources, recipients, and transporters are essential to ensure the safety of food supply (Levinson 2009).

Food safety hazards may occur at a variety of points in the food chain. Therefore, food safety is a responsibility that is shared by producers, processors, distributors, retailers, and consumers. An important preventative approach that may be applied at all stages in the food chain involves Hazard Analysis Critical Control Point (HACCP) system (FAO and WHO 2003). HACCP is the internationally recognized reference system for food safety management, and based on seven key principles, which is widely used worldwide. The seven principles are:

(1)   identify hazards, assess risk, and list controls;
(2)   determine critical control points;
(3)   specify criteria to ensure control;
(4)   establish monitoring system for control points;
(5)   take corrective action whenever monitoring indicates criteria are not met;
(6)   verify that the system is working as planned; and
(7)   keep suitable records (NZFSA 2003).

Quality is defined by ISO as "the totality of features and characteristics of a product that bear on its ability to satisfy stated or implied needs" (Van Reeuwijk 1998). Also, quality can be defined as "conformance to requirement", "fitness for use" or, more appropriately for foodstuffs, "fitness for consumption". Thus, quality can be described as the requirements necessary to satisfy the needs and expectations of the consumer (Ho 1994; Peri 2006). However, food quality is very general, implying many expectations which can be different from consumer to consumer. Quality includes attributes that influence a product's value to the consumer. Quality does not refer solely to the properties of the food itself, but also to the ways in which those properties have been achieved (Morris and Young 2000). The categories of quality attributes are listed in Table 3.2.

Many experts have argued that safety is the most important component of quality since a lack of safety can result in serious injury and even death for the consumer. Safety differs from many other quality attributes since it is a quality attribute that is difficult to observe. A product can appear to be of high quality (i.e., well colored, appetizing, and flavorful, etc.), but it can be unsafe because it might be contaminated

Table 3.2 Categories of food quality attributes

External	Internal	Experience	Hidden
Appearance (sight)	Smell (aroma)	Taste (flavor)	Wholesomeness
Feel (touch)	Size (inner)	Freshness	Nutritive value
Defects	Texture	Tenderness	Safety
Package and label	Color (inner)	Juiciness	Process/inputs

with undetected pathogenic organisms, toxic chemicals, or physical hazards (UN 2007). Rohr et al. (2005), Grunert (2005) and Pinto et al. (2006) agreed that food safety has become an important food quality attribute.

Defects and improper food quality may result in consumer rejection and lower sales, while food safety hazards may be hidden and go undetected until the product has been consumed. If detected, serious food safety hazards may result in market access exclusion and major economic loss and costs. Since food safety hazards directly affect public health and economies, achieving proper food safety must always take precedence over achieving high levels of other quality attributes (UN 2007). These two have obvious links, but food quality is primarily an economical issue decided by the consumer, while the food safety is a governmental commitment to ensure that the food supply is safe for consumers and meet regulatory requirements (Sarig 2003). Quality is seen to lead to taste, health, safety, and pleasure. Similarly, safety is seen to be the consequence of controlling origin, best before date and quality, while resulting in health and a feeling of calm. Both quality and safety are interrelated and linked to trust/confidence (Rijswijk and Frewer 2006).

Food quality, especially food safety, is a major concern faced in food industry today. The production and consumption of food is central to any society and has a wide range of social, economic, and, in many cases, environmental consequences.

3.3 Medicinal Cold Chain

Following manufacture, some medicinal products need to be stored and shipped at lower than ambient temperatures to assure their quality and efficacy. Medical products like vaccines, insulin, blood, and some pharmaceuticals are temperature sensitive and can be classified as high risk because they are at risk from freezing as well as elevated temperatures (Todd 2008). Therefore, they need a cold chain logistics that monitor and control temperature in order to prevent damage caused by heat exposure and to maintain quality and safety in every stage of the supply chain (Ruiz-Garcia and Lunadei 2010).

Global regulatory requirements for the handling including packaging, storage, and distribution of thermally liable pharmaceutical products have emphasized the importance of assuring that product quality and integrity are not compromised in the distribution channel. Due to the presence of multiple uncontrolled variables in the distribution process, developing an appropriate monitoring program is essential (Bishara 2006).

Typically, the distribution chain for medicinal products is complex, potentially involving several storage locations, wholesalers, and modes of transport, before the delivery finally reaches the pharmacy. Packaging plays a key role in protecting the product from damage during its passage from the producer to the consumer. Therefore, cold chain products should be packed in such a way as to ensure that the required temperatures are maintained throughout the journey and the medicines are transported in accordance with their labeling requirements not to jeopardize their quality (Todd 2008). United Parcel Service (UPS) has various service for medical cold chain such as UPS Temperature True, UPS Temperature True Cryo, UPS Temperature True Packaging, PharmaPort 360, UPS Proactive Response, and UPS Proactive Response Secure (http://www.ups.com). Among services, Temperature True Packaging provides prequalified packaging systems for customers' temperature-sensitive products to protect products with a range of time and temperature requirements.

FedEx and DHL have made significant investments in their healthcare supply chains as well. They also see big potential in medicinal cold chain (http://www.cnbc.com/2015/10/02/ups-fedex-and-dhl-bet-big-on-health-care-logistics.html).

According to IATA (2014), the global pharmaceutical industry spent about $8.36 billion on cold chain logistics in 2014 and was expected to expand to more than $10 billion by 2018. A basic pharmaceutical supply chain can be seen in the following Fig. 3.4.

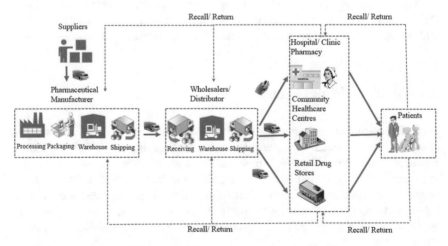

Fig. 3.4 Basic pharmaceutical supply chain

3.4   Vaccine Cold Chain

Immunization is recognized as one of the most successful public health programs saving millions of lives every year from contagious diseases that is implemented through vaccination—the process of injecting vaccine to produce immunity that can protect against a disease. Between 2010 and 2018, 23 million deaths were averted with measles vaccine alone. More than 20 life-threatening diseases can now be prevented by immunization. There are now vaccines to protect against malaria, dengue, and Ebola virus disease, and promising vaccines against respiratory syncytial virus, tuberculosis, and all influenza virus strains are in the pipeline. New research on broadly neutralizing antibodies and therapeutic vaccines is opening fresh horizons. Increasingly, vaccines are protecting health beyond infancy—in adolescence and adulthood, during pregnancy and for older people (WHO 2020).

Vaccines are vital to the prevention and control of many infectious disease outbreaks. Moreover, they are widely seen as critical for addressing emerging infectious diseases, for example by containing or limiting outbreaks of infectious diseases or combatting the spread of antimicrobial resistance. Regional outbreaks (e.g., of Ebola virus disease), the COVID-19 pandemic and the threat of future pandemics (such as with a novel flu strain) highlighted that we still have challenges to build resilient health systems.

Delivering vaccines to all corners of the world is a complex task as it requires a system of refrigeration to store, manage, and transport from the time they are manufactured until the moment of vaccination. The system for distributing vaccines in a potent state from the manufacturers to the actual vaccination sites is called the cold chain. The vaccine cold chain was deliberately separated from other medical distribution systems to assure timely access to and proper control of vaccines and injection materials (Lloyd and Cheyne 2017).

All vaccines are sensitive biological substances which are susceptible to heat, light, and/or freezing. Exposure of vaccine to temperatures below or above the recommended ranges in the cold chain may decrease or lost vaccine potency. Once the vaccines potency is lost, it cannot be regained or restored and the vaccine will no longer provide any protection against the target disease leading to financial loss and risk of contracting vaccine preventable illness. The risk of cold chain failure increases as the vaccines move along the cold chain from the manufacturer to the vaccine recipient and is greatest at the vaccinator level.

The stability of vaccines' potency compromises if they are stored in unfavorable conditions. World Health Organization (WHO)'s vaccine management handbook (WHO 2015a, b) and MOH Singapore's guideline (MOH Singapore 2005) stated vaccines' sensitivity to three types of condition.

3.4.1   Sensitivity to Heat

Although all vaccines are sensitive to heat, some vaccines are more sensitive to heat than others. Vaccines do not change their appearance when potency is lost. A complete laboratory test is the only means to assess whether a vaccine in a vial has lost its potency. The following vaccines are listed in order of heat sensitivity from most heat sensitive to least sensitive.

- Oral Polio Vaccine (OPV);
- Measles (lyophilised);
- Diphtheria, Tetanus and Pertussis (DTP);
- Yellow Fever;
- Bacillus Calmette–Guérin (BCG);
- *Haemophilus influenzae* Type b (Hib);
- Diphtheria and Tetanus toxoid (DT);
- Tetanus and Diphtheria toxoid (Td);
- Tetanus toxoid (TT), Hepatitis B.

3.4.2   Sensitivity to Freezing

Some vaccines are also sensitive to extreme cold. For these vaccines, freezing or exposure to temperatures below zero degrees Celsius can also cause loss in potency and render the vaccines useless. Although there are some vaccines that can tolerate freezing such as freeze-dried vaccines (e.g., Measles, OPV), freezing can irreversibly damage vaccines that contain aluminum-salt adjuvants. For these vaccines, it is therefore essential to protect them not only from heat, but also from freezing. The vaccines sensitive to freezing (as well as to heat) from top most sensitive to least sensitive are:

- Hepatitis B;
- HiB (liquid);
- DTP;
- TD;
- TT.

3.4.3   Sensitivity to Light

Some vaccines are also very sensitive to strong light. For these vaccines, exposure to ultraviolet light will cause loss of potency, so they must always be protected against sunlight or fluorescent (neon) light. BCG, Measles, Measles and Rubella (MR), Measles, Mumps and Rubella (MMR) and Rubella vaccines are sensitive to light (as well as to heat). Normally, these vaccines are supplied in vials made from dark brown glass, which gives them some protection against damage from light. However, care must still be taken to keep them covered and protected from strong light at all times. They should not be stored in a cooler with a glass door and should preferably be stored in the dark.

3.4.4 Recommended Storage Temperature for Vaccines

With a few exceptions, majority of the time, temperatures were kept between recommended temperature range of 2–8 °C. However, it is important to always refer to the manufacturer's storage recommendation as the recommendations could be significantly different depending on the types of vaccine. Some vaccines are lyophilized and must be combined with a liquid component called a diluent before injection; a process called reconstitution. All diluents must be protected from freezing. In the absence of specific instructions, these diluents should be treated the same as any other non-cold chain pharmaceutical, protected from freezing, and stored and transported between +2 and +25 °C. Freeze-dried vaccines and their diluents should always be distributed together in matching quantities. The vaccines must be always kept in the cold chain between +2 and +8 °C or, optionally, at −15 to −25 °C if cold chain space permits.

There are several types of vaccines, including:

- inactivated vaccines (use the killed version of the germ that causes a disease);
- live-attenuated vaccines (use a weakened (or attenuated) form of the germ that causes a disease);
- messenger RNA (mRNA) vaccines (make proteins in order to trigger an immune response);
- subunit, recombinant, polysaccharide, and conjugate vaccines (use specific pieces of the germ—like its protein, sugar, or capsid (a casing around the germ));
- toxoid vaccines (use a toxin (harmful product) made by the germ that causes a disease);
- viral vector vaccines (use a modified version of a different virus as a vector to deliver protection).

(Source: https://www.hhs.gov/immunization/basics/types/index.html).

Some COVID-19 vaccines using to fight for COVID-19 global pandemic are stated in Table 3.3 with their types and storage temperature requirement.

3.4.5 Cold Chain Equipment for Storage/Distribution

To be effective, cold chain equipment must be properly maintained. The usual types of cold chain equipment are:

Table 3.3 COVID-19 vaccine types and storage temperatures

Vaccine	Vaccine types	Storage temperatures (°C)
Pfizer BNT162	mRNA vaccine	−60 and −80 °C
Moderna mRNA-1273	mRNA vaccine	−15 and −25 °C
Oxford–AstraZeneca	Viral vector vaccine	2 and 8 °C
Johnson & Johnson	Viral vector vaccine	2 and 8 °C
Sputnik V	Viral vector vaccine	2–8 °C/−18 °C (freeze-dried)
CoronaVac Sinovac	Inactivated virus vaccine	2 and 8 °C
Novavax	Protein subunit vaccine	2 and 8 °C

- refrigerator/freezer;
- thermometer;
- temperature data loggers;
- cold box;
- vaccine carrier (e.g., thermos flask).

Purpose-build refrigerators are the best and standard for storing large inventory of vaccine and biologics except few conditional uses of domestic refrigerators. In areas with unreliable electricity, ice-lined refrigerators have been utilized to provide cooling during extended power cuts. Immunization programs need to cover not only urban areas but also remote and off-grid rural areas. Especially, operating cold chain equipment in remote rural areas without grid electricity has presented a serious challenge for unbroken cold chain. WHO and UNICEF recommend using solar refrigerators in off-grid locations with sufficient solar irradiance. The common types of solar refrigerator are battery-powered and direct-drive solar refrigerator. Figure 3.5 shows the image of direct-drive solar-powered vaccine refrigerator of B-medical systems (https://www.bmedicalsystems.com/) for rural immunization programs (WHO 2015a).

Cold boxes and vaccine carriers are insulated containers that can be lined with coolant packs and keep vaccines and diluents cold during transportation and/or short period storage. Vaccine carriers are smaller than cold boxes and are easier to carry when walking, therefore suitable for use by health workers during immunization campaigns and out-reach services. As these are passive devices, coolant packs are accessories for both cold boxes and vaccine carriers providing the cooling energy/warming for a limited time period. Figure 3.6 shows the images of cold box and vaccine carriers used in vaccine cold chain.

Fig. 3.5 Direct-drive
solar-powered vaccine
refrigerator (Reprinted by
permission of B-medical
systems)

Fig. 3.6 Cold box, vaccine
carrier, and ice packs

3.4.6   Temperature Monitoring for Vaccine Supply Chain

In order to maintain vaccine quality, it is essential to monitor the temperature of vaccines throughout the supply chain. Effective monitoring and record-keeping achieves the following objectives:

- verification that vaccine storage temperatures are within the acceptable ranges of +2 to +8 °C in cold rooms and vaccine refrigerators and −25 to −15 °C in freezer rooms and vaccine freezers;
- detection of out-of-range storage temperatures so that corrective action can be taken;
- detection of out-of-range transport temperatures so that corrective action can be taken.

WHO recommends temperature monitoring devices based on the specific cold chain equipment application and the intended purpose of monitoring. Temperature monitoring technology is evolving rapidly from traditional thermometer to chemical indicators and digital data loggers. A sample of digital thermometer used for vaccine can be seen in Fig. 3.7. The temperature monitoring devices used in district, primary and subnational, national and international vaccine shipments, and storage facilities are:

- vaccine viral monitors (VVMs);
- cold chain monitor (CCM) cards;

Fig. 3.7 Digital thermometer used for vaccines

- electronic shipping indicators;
- programmable electronic temperature and event logger systems;
- integrated digital thermometers;
- 30-day electronic temperature recorders (30 DTRs);
- electronic freeze indicators;
- stem thermometers.

3.4.7 Vaccine Viral Monitors (VVMs)

VVMs are the only temperature monitoring devices that routinely accompany vaccines throughout the entire supply chain. A VVM is a chemical indicator label which is applied to a vaccine vial, ampoule or other type of primary container by the vaccine manufacturer. As the container moves through the supply chain, the VVM records cumulative heat exposure through a gradual change in color (Fig. 3.8). VVMs are included on nearly all vaccines purchased through the United Nations Children's fund (UNICEF).

The main purpose of VVMs is to ensure that heat-damaged vaccines are not administered. VVM status is also used to decide which vaccines can safely be kept after a cold chain break occurs; this minimizes unnecessary vaccine wastage. In addition, VVM status helps determine the order in which vaccines should be used—a batch of vaccine with VVMs that show significant heat exposure but have not yet reached their discard points should be distributed and used ahead of a batch that shows lower heat exposure, even if the expiry date is later.

In order to ensure vaccine traceability in the supply chain, VVM status should be checked before dispatch and manually recorded on arrival vouchers at stores and health facilities. VVMs should also be checked by health workers before the vaccine is used. There are currently four VVM types. In liaison with WHO, vaccine manufacturers use the type which is most appropriate to the stability profile of their

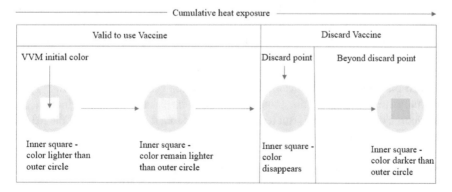

Fig. 3.8 Vaccine viral monitor color change sequence

vaccine. The detail guidelines on how to monitor temperatures in the vaccine supply chain is stated in WHO's Vaccine Management Handbook: Module VMH-E2 (WHO 2015b).

3.5 Socio-economic and Environmental Impacts of Cold Chain

3.5.1 Social Impacts

Temperature abuse in the food cold chain can make microbial growth and spoilage of food so become the factor causing food-borne illness. Food safety is an increasingly important public health issue. Outbreaks of foodborne illness can damage trade and tourism and lead to loss of earnings, unemployment, and litigation (CAC 2003). Globally, the incidence of foodborne diseases is increasing and international food trade is disrupted by frequent disputes over food safety and quality requirements (FAO 2003). Unsafe food causes many acute and lifelong diseases, ranging from diarrhoeal diseases to various forms of cancer.

World Health Organisation (WHO 2002) estimated that foodborne and water-borne diarrhoeal diseases taken together killed about 2.2 million people annually, 1.9 million of them children. In industrialized countries, the percentage of the population suffering from foodborne diseases each year has been reported to be up to 30%. In United States (US), for example, around 76 million cases of foodborne diseases, resulting in 325,000 hospitalizations and 5000 deaths, are estimated to occur each year. The high prevalence of diarrhoeal diseases in many developing countries highlights major underlying food safety problems (WHO 2007).

3.5.2 Economic Impacts

WHO (2002) stated that foodborne diseases not only significantly affect people's health and well-being, but they also have economic consequences for individuals, families, communities, businesses, and countries. These diseases impose a substantial burden on healthcare systems and markedly reduce economic productivity. There are only limited data on the economic consequences of food contamination and foodborne disease. In 1995, studies in US reported that the annual cost of the 3.3–12 million cases of foodborne illness caused by seven pathogens was approximately US\$ 6.5–35 billion. Recently, former US Food and Drug Administration (FDA) economist Robert L. Scharff estimated the total economic impact of foodborne illness across the nation to be a combined \$152 billion annually (Scharff 2010).

US Department of Agriculture (USDA) estimates cost of illness associated with medical expenses and losses in productivity from five major types of foodborne illnesses at \$6.9 billion annually (Vogt 2005). In European Union, annual costs leveled on the healthcare system as a consequence of *Salmonella* infections are estimated to be around 3 billion euros (Asian Productivity Organisation 2009). The medical costs and the value of the lives lost during just five foodborne outbreaks in England and Wales in 1996 were estimated at UK£ 300–700 million. The cost of the estimated 11,500 daily cases of food poisoning in Australia was calculated at AU\$ 2.6 billion annually. The increased incidence of foodborne disease due to microbiological hazards is the result of a multiplicity of factors, all associated with our fast-changing world (WHO 2002).

3.5.3 Environmental Impacts

With the growth of international food trade, environmental impact of food supply chain has become a growing concern. The distance that food travels from the farm where it is produced onto the kitchen in which it is consumed is longer than before. Therefore, the use of energy, resources, and emission of greenhouse gases (GHG) in the entire food cycle, including production, consumption, and transport is unavoidable. The initiatives to use carbon labeling (i.e., carbon footprints of the products) and conception of food miles (the distance that food is transported as it travels from producer to consumer) indicate that the food chain needs more environmental-friendly solution to reduce the environmental impacts such as pollution and global warming.

In many countries, one of the problems concerned with food safety and quality is food spoilage. Food spoilage is wasteful, costly, and can adversely affect trade and consumer confidence. Naturally, all foods have limited lifetime and most foods are perishable. Safe and high-quality chilled foods require minimal contamination during manufacture, rapid chilling, and temperature control along the chain (Martin and Ronan 2000).

The International Institute of Refrigeration indicates that about 300 million tons of produce are wasted annually through deficient refrigeration worldwide. In US, the food industry annually discards USD 35 billion worth of spoiled goods. The wastage of food and resources used for growing unused products is also a big issue for the environment (Flores and Tanner 2008).

UK households waste 6.7 million tons of food every year. Waste Resources and Action Programme estimates that a third of the food bought is uneaten and thrown out. If that food waste was eradicated, it would be equivalent to taking one in five cars off the road. Every tons of food waste is responsible for 4.5 tons of carbon dioxide. The food waste which is thrown as landfill where it is liable to create methane, a powerful greenhouse gas which is over 20 times more potent than carbon dioxide makes a significant environmental impact (WRAP 2008). Research by Australia Institute indicates that Australians throw away about \$5.2 billion worth of food every year. Wasting food also wastes the water that went into its production (Baker et al. 2009).

3.6   Research Trends of Cold Chain

In food supply chain, many kinds of products must be handled under controlled environmental parameters, such as temperature, humidity, vibrations, and light exposure due to their impacts on quality and safety of food. Among all these parameters, the temperature is usually one of the major concerns due to its huge effects. If the temperature of some chilly foods exceeds specific limits, the rise in temperature of just a few degrees can cause microbial growth leading to the great decrease of quality and the increase of the risk of food poisoning (Carullo et al. 2008). Perishable food products must be continuously monitored for safety and quality concerns throughout the whole supply chain. A breakdown in temperature control at any stage will impact on the final quality of the product (SARDI 2006).

Control over physical movement of goods is very common in typical supply chain, but quality control and monitoring of goods in cold chain is an increasing concern for producers, suppliers, transport decision makers, and consumers. Commercial systems are currently available for monitoring containers, refrigerated chambers, and trucks, but they do not give complete information about the cargo, because they typically measure only a single or very limited number of points. It shows how important it is to have the system that can monitor and maintain the temperature and enable visibility not only the physical flow but also the quality condition throughout the whole cold chain and become an interesting research area in cold chain.

Certainly, technology plays a more significant role in cold chain. The development of telecommunication, information technology and information system, especially the rise of wireless sensing technologies, such as RFID and WSN, provides a feasible way to monitor the safety and quality of food cold chain. Integrating RFID systems with condition monitoring systems will enhance existing track and trace applications, not only the location and movement history, but also the condition of perishable products. Moreover, the availability of product trace history data in combination with historical condition monitoring data can facilitate numerous decision-making processes (Mitsugi et al. 2007; Óskarsdóttir and Oddsson 2019).

WSN provides a flexible and powerful solution to monitor the cold chain in every stage of the logistical chain. WSN can monitor the whole supply chain in real time starting from the firm to the consumer's plate. Refrigerated vehicles can be a crucial point as the products may undergo transient conditions during transport and distribution process while the temperature is easily controlled and monitored in the factory plants and warehouses. Therefore, convenient and reliable monitoring systems are increasingly demanded for vehicle refrigerators. Condition monitoring sensors can be classified into two categories: continuous monitoring sensor and discrete (event) monitoring (Bowman et al. 2009). The supply chain, especially cold chain, needs continuous sensing to monitor the abnormal changes in environmentally sensitive products and to provide intelligent responses in the events of failure to keep optimal conditions. The example of continuous sensing in a supply chain is illustrated in Fig. 3.9. In order to monitor product condition continually, efficient power management is a critical factor.

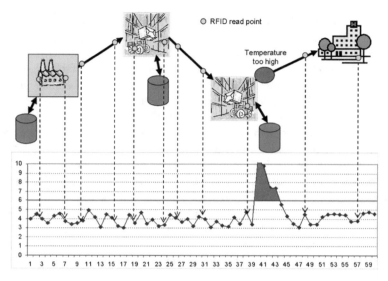

Fig. 3.9 Continuous sensing in the supply chain (Reprinted by permission of www.bridge-projec t.eu—Bowman et al. 2009)

In fact, cold chain refers to the equipment and processes employed to ensure the temperature preservation of perishables and other temperature-controlled products from the production to the consumption end in a safe, wholesome, and good quality state. Therefore, facilities and processes may determine the success of the cold chain. A previous research article indicate that the refrigerated vehicles can be a crucial point as the products may undergo transient conditions during transport and distribution process while the temperature is easily controlled and monitored in the factory plants and warehouses. A study showed that temperature-controlled shipment rises above the specified temperature in 30% of trips from the supplier to the distribution center and in 15% of trips from the distribution center to the store. Lower than required temperatures occur in 19% of trips from supplier to distribution center and in 36% of trips from the distribution center to the store (White 2007).

The uses of application-specific sensor designs are found in Carullo et al. (2008) where they proposed an architecture, which is based on the specifically designed sensor nodes that are inserted into the products to monitor the thermal behavior (air and product temperature) of packed goods in the containers for cold chain assurance. The functional tests and the experimental tests of the proposed system on a refriger-ated vehicle during a distribution are reported. For wine logistics chain, Mattoli et al. (2010) developed a Flexible Tag Data logger (FTD) that can be attached to the bottles to collect environmental data (light, humidity, and temperature) in order to trace the wine bottles that leave the producer cellar for the transport to a shop. The history data stored in the FTD can be read by smart phone or Personal Digital Assistant with integrated infrared port to evaluate whether the bottle of wine has been maintained in a safe manner or not.

Jedermann et al. (2006) described an autonomous and integrated sensor system for ICs combining WSNs, RFID, and software agents to monitor fruit transports. They proposed a miniaturized high-resolution gas chromatography system for measuring ripening indicator ethylene. Again, miniaturized RFID temperature loggers are applied to analyze the amount of local deviations, detect temperature gradients, and estimate the minimum number of sensors that are necessary for reliable monitoring inside a truck or container (Jedermann et al. 2009). Ruiz-Garcia et al. (2007) analyzed a monitoring of intermodal refrigerated fruit transport that integrates WSNs with multiplexed communications, fleet management systems, and mobile networks. They also studied the feasibility and performance of ZigBee motes for monitoring the refrigerated conditions in fruit chambers with low temperatures, high humidity, and different cargo densities. For WSN, ZigBee protocol is considered as the best candidate as it satisfies reliability, low cost, and low power consumption as well as multi-hop communication (Ruiz-Garcia et al. 2008). Shan et al. (2004, 2005) investigated the future intelligent WSN, which integrates artificial intelligent technologies such as neural network, fuzzy logic, and neuro-fuzzy, and developed a prototype system capable of monitoring several parameters relating to vehicle refrigerators such as air temperature and door status in real time.

Traceability systems are considered important in supply chains today to ensure the safety of the product. The most common types of traceability technologies that capture data in the food SC are paper records, barcodes, RFID, and WSNs (Óskarsdóttir and Oddsson 2019). Abad et al. (2009) presented a real-time traceability and cold chain monitoring system for foods by using smart RFID tag in intercontinental fresh fish logistic chain. They proved that the prototyped semi-passive smart tags with integrated light, temperature, and humidity sensors achieve important advantages over traditional data loggers in terms of storage, reusability, automation, visibility, concurrency, and durability. Wang et al. (2010) presented a real-time monitoring and decision support system, with the combination of existing technologies such as RFID, WSN, GPS, and rule-based decisions to improve the delivery system for perishable products. Based on the mathematical models, and data from RFID and the sensor network, the quality of the goods can be predicted by the forecast module. Rule-based decision module will provide the user with suggestions on how to cope with the abnormality.

These days blockchain technology has gained global attention with potential to revolutionize supply chain management and sustainability achievements. Blockchain is a shared, distributed, and immutable ledger system that facilitates recording transactions and tracking assets in a business network. Because of its reliability, transparency, and immutability, it can be applied to many functions of SCM such as logistics, quality assurance, inventory management, and forecasting (Song et al. 2019; Kouhizadeh et al. 2021). Since blockchain works together with other technologies (e.g., IoT, RFID, GPS, etc.), Tsang et al. (2019) proposed a blockchain–IoT-based food traceability system that integrates blockchain, IoT technology, and fuzzy logic to build a total traceability shelf life management system for managing perishable food. Blockchain can improve supply chains by enabling faster and more cost-efficient delivery of products, enhancing products' traceability, improving coordination between supply chain partners, enabling secure transactions with data management, and aiding access to financing. The recent research interest shows that the application of blockchain in traceability, especially for food and agricultural supply chains (Mirabelli and Solina 2019; Song et al. 2019; Kamilaris et al. 2019).

Another trend is the global or local connectivity of cold chain application as they use web-based interface that allows configuration of the network and access to real-time and archived temperature data. The availability of real-time information of the products allows proactive measure to do when the quality and safety problems arise. Crowley et al. (2004) described the implementation of a WSN for the temperature monitoring of shellfish catches over the Internet. Temperature loggers are used in shellfish boxes, information is transferred to a server via the GSM network which allows for functional querying of real-time and archived temperature information. There is an increase in research considering the prospect of implementing digital technologies like IoT in the supply chain. Ruan and Shi (2016) proposed IoT-based framework for fruit e-commerce delivery. Within the framework, IoT-related technologies such as GPS, RFID, and WSN are applied using mobile communication networks and the Internet to enable real-time monitoring and assessment of freshness of the product in online demand fulfilments.

The importance in cold chain is to establish the process to maintain product quality and ensure safety complying with regulations and industry standards for temperature-sensitive products. In Bogataj et al. (2005), a mathematical model for cold traceability is presented for enabling visibility to the quality of products tracking deterioration along supply chain to keep the product on the required level of quality and quantity at the final delivery. Rong et al. (2009) presented a mixed-integer linear programming model for the planning of food production and distribution with a focus on product quality, which is strongly related to temperature control throughout the supply chain. The approach combines decision-making on traditional logistical issues such as production volumes and transportation flows with decisions on storage and transportation temperatures.

To be an unbroken cold chain, not just the temperature-controlled system, the thermal packaging that is designed in consideration with cooling kinetics is also important. Wu et al. (2019) evaluated three packaging designs for citrus fruit and found that Supervent packaging outperformed than the Standard and Opentop packaging as it allows fastest and most uniform cooling. Defraey et al. (2015) highlighted that there are many important factors to consider in packaging design such as product cooling rate, box ventilation, product quality and shelf life, box mechanical strength, and energy consumption of the ventilation system. Modified Atmosphere Packaging (MAP) is used commercially to extend the shelf life of fruits. Joshi et al. (2019) studied MAP for strawberry by building a mathematical model and found interventions are needed to control quality, and reduction of respiration rate and water transfer is effective means to control weight loss or spoilage occurred to fruits. The result shows that the attainment of desired gas (O_2, CO_2) concentrations inside MAP relies on the product respiration and the mass transfer through packaging and will affect the quality.

A summary of some of the commercial cold chain solutions (https://www.sensit ech.com/en/, https://www.ellab.com/, https://www.deltatrak.com/, https://www.cry opak.com/, https://www.elpro.com/en/) is shown in Table 3.4.

In summary, it is important to recognize the fact that technology plays a key role in improving the global cold chain. Instead of using single technology, convergence of technologies is the trend complementing each other to form more resilient, intelligent, and powerful systems. With COVID-19 and global regulatory framework constantly changing, technology will expand its critical role in visibility, monitoring, and compliance going forward. The need for tracing food sources and monitoring environmental conditions of food cargo events will be elevated since inspection processes will be streamlined as regulatory compliance increases. In addition, as transit times increase due to the supply chain challenges connected to COVID-19, controlled-atmosphere refrigerated technology will become more necessary. Transportation is the most focused area of research in cold chain as it is a weak link in cold chain due to its mobility and exposure to weather conditions. Different application-specific sensors are designed to use for complex food supply chain since it has many diverse process and products. Global connectivity can be achieved by using web-based IoT systems and blockchain, and this is the promising trend of cold chain system today to trace back the quality and safety of food they bought and to securely connect among actors. Most of the previous research mentioned above focused on condition monitoring, device management, and traceability area but not much focused on product characteristics and optimal setting for monitoring environment. Mercier et al. (2017) suggested the future research on the cold chain should pay more attention on three critical areas: efficient precooling, cold chain management system, and cold chain in developing countries. Although the industry today still uses TTI and data loggers commercially for perishable products, the adoption of more efficient and smarter tech-based solutions is expected to see in very near future.

Table 3.4 Commercial cold chain solutions

Suppliers	Solutions	Features	Benefits
Sensitech (US)	ColdStream Select, ColdStream Site, ColdStreamWatch	• Use traceable calibrated TempTale® Ultra-RF-enabled data loggers, • Enables real-time, end-to-end visibility creating a comprehensive IoT ecosystem, • With on-demand reports and dashboard for analysis, • Reliable internal and external data gathering and integration for food, life sciences, consumer, and industrial	• Improve freshness, • Reduce shrink, • Increase profits of perishables, • Achieve higher levels of customer satisfaction
Hanwell (Ellab) (Denmark)	Hanwell Pro (Superior) Hanwell Icespy (premium) Notion Lite (plug and play)	• Use radio or GPRS-enabled temperature data loggers, • Fully automated alarms and automated reporting, • Option with hardwares and Hanwell EMS for seamless access from anywhere • Interactive graphs and data analysis accessible in smart phones	• Integrated and highly scalable solutions for different types of cold chain • A range of wireless data loggers available for a vast range of measurement types
DeltaTRAK . (US)	FlashTrak	• Use FlashLink IoT data loggers, • Web-based application for retrieving, analyzing, and sharing temperature data, • Enables viewing of trip data and graphs • Location data plotted on Google Maps • Use chart recorder for preventive maintenance, • FlashTrak Cloud Services	• Live data visibility • Traceability • Compliance for all modes of transportation • BI and analytics • Both physical and digital monitoring available to ensure integrity
Cryopak (Canada)	Escort Data Loggers	• Real-time alarm call to a phone, computer, and e-mail, • Faster upload and download, • Programmable loggers, • Clear visual graphs	• Data secured with digital signatures • Both wireless and web-based solution available

(continued)

Table 3.4 (continued)

Suppliers	Solutions	Features	Benefits
ELPRO (Swiss)	LIBER ITS (Electronics Smart Indicators) LIBERO CS, CB, and CD (Data Loggers) LIBERO Gx (Real-Time Data Loggers)	• Electronic multi-level indicator • PDF reports Bluetooth or USB data loggers • Measurable from −200 °C to + 200 °C Measure many sensor parameters • Combine real time (IoT) with traditional data loggers • Can use for 1 trip (110 days) or 1 years • Cloud-based database	• IATA compliant • Wide range of solutions available based on products • Enable supply chain visibility • Mobile app and Manager software for process automation and analytics • Wired or wireless options

References

Abad E, Palacio F, Nuin M, Zárate GD, Juarros A, Gómez JM, Marco S (2009) RFID smart tag for traceability and cold chain monitoring of foods: demonstration in an intercontinental fresh fish logistic chain. J Food Eng 93(4):394–399

Asian Productivity Organisation (2009) Food safety management manual. Tokyo, Japan

Baker D, Fear J, Denniss R (2009) Analysis of household expenditure on food. Policy brief No 7

Beasley SD (2002) Helping exports keep their cool. AgExporter 14(4):4–5

Bharti A (2017) Recent trends in cold chain management. J Manag Sci Technol 4(2):41–52. ISSN 2347-5005

Bishara RH (2006) Cold chain management—an essential component of the global pharmaceutical supply chain. Am Pharmaceut Rev (Jan/Feb). Available online via http://intelsius.com/wp-content/uploads/2011/10/Pharma-Cold-Chain-Bishara_APR.pdf. Accessed 20 Apr 2021

Bogataj M, Bogataj L, Vodopivec R (2005) Stability of perishable goods in cold logistics chains. Int J Prod Econ 93–94:345–356

Bourlakis MA, Weighman PWH (2001) Introduction to the UK food supply chain. In: Bourlakis MA, Weighman PWH (eds) Food supply chain management. Blackwell Publishing, pp 179–198

Bowman AP, Ng J, Harrison M, Lopez TS, Illic A (2009) Sensor based condition monitoring. Bridge Project. www.bridge-project.eu

CAC (2003) Basic texts on food hygiene, 3rd edn. Codex Alimentarious Commission. Retrieved from http://www.fao.org/3/y5307e/y5307e00.htm. Accessed 25 Apr 2020

Carullo A, Corbellini S, Parvis M, Reyneri L, Vallan A (2008) A measuring system for the assurance of the cold-chain integrity. In: Proceedings of the IEEE international instrumentation and measurement technology conference, Vancouver, Canada, pp 1598–1602

CHGE (2012) What is the food supply chain? Center for Health and Global Environment, Harvard University. Retrieved from https://hwpi.harvard.edu/files/chge/files/lesson_4_1.pdf. Accessed 25 Apr 2021

Cleland AC (1996) Package design for refrigerated for refrigerated food: the need for multidisciplinary project teams. Trends Food Sci Technol 7:269–271

Coulomb D (2008) Refrigeration and cold chain serving the global food industry and creating a better future: two key IIR challenges for improved health and environment. Trends Food Sci Technol 19:413–417

Crowley K, Frisby J, Edwards S, Murphy S, Roantree M, Diamond D (2004) Wireless temperature logging technology for the fishing industry. In: IEEE Sensors 2004, Vienna, Austria, pp 571–574

Defraeye T, Cronjé P, Berry T, Opara UL, East A, Hertog M, Vervoven P, Nicolai B (2015) Towards integrated performance evaluation of future packaging for fresh produce in the cold chain. Trends Food Sci Technol 44:201–225

Dittmer P, Veigt M, Scholz-Reiter B, Heidmann N, Paul S (2012) The intelligent container as a part of the internet of things, a framework for quality-driven distribution for perishables. In: IEEE international conference on Cyber technology in automation, control, and intelligent systems (CYBER), 27–31 May 2012, pp 209–214, Bangkok

FAO (2003) FAO's strategy for a food chain approach to food safety and quality: a framework document for the development of future strategic direction. Retrieved from http://www.fao.org/DOCREP/MEETING/006/y8350e.HTM. Accessed 28 Aug 2020

FAO & WHO (2003) Assuring food safety and quality: guideline for strengthening national food control system. Joint FAO/WHO Publication. Retrieved from http://www.fao.org/3/a-y8705e.pdf. Accessed 25 Apr 2019

Fernie J, Sparks L (2004) Logistics and retail management: insights into current practice and trends from leading experts, 2nd edn. Kogan Page Limited, UK and US

Flores SE, Tanner D (2008) RFID technologies for cold chain applications. Int Inst Refriger Bull 15(4):4–9

Fu B, Labuza TP (1997) Shelf-life testing: procedures and prediction methods. In: Erickson MC, Hung YC (eds) Quality in frozen food. Springer, Boston

Grunert KG (2005) Food quality and safety: consumer perception and demand. Eur Rev Agric Econ 132(3):369–391

Heap RD (2006) Cold chain performance issues now and in the future. IRHACE J

Ho SKM (1994) Is the ISO 9000 series for total quality management? Int J Q Reliab Manag 11(9):74–89

IATA (2014) News brief: new global certification program for handling cold-chain pharmaceuticals. Retrieved from http://www.iata.org/pressroom/pr/Pages/2014-08-28-01.aspx. Accessed 25 Apr 2019

IGD (2001) Retail logistics 2001. Institute of Grocery Distribution (IGD) Business Publication, Lechtmore Heath

IIR-UNEP (2007) Refrigeration drives sustainable development. International Institute of Refrigeration. United Nations Environment Programme. Retrieved from http://www.iifiir.org/userfiles/file/_Backup/Dossiers_Exclusifs/RDSD_EN.pdf. Accessed 25 Apr 2019

James SJ, James C, Evans JA (2006) Modelling of food transportation systems—a review. Int J Refrig 29:947–957

Jedermann R, Behrens C, Westphal D, Lang W (2006) Applying autonomous sensor systems in logistics combining sensor networks. Rfids Softw Agents Sens Actuat Phys 132(1):370–375

Jedermann R, Edmond JP, Lang W (2007) Shelf life prediction by intelligent RFID. In: Haasis HD, Kreowski HJ, Scholz-Reiter B (eds) Dynamics in logistics—first international conference, LDIC2007 Bremen. Springer, Berlin, pp 231–238

Jedermann R, Ruiz-Garcia L, Lang W (2009) Spatial temperature profiling by semi-passive RFID loggers for perishable food transportation. Comput Electron Agric 65:145–154

Joshi K, Tiwari B, Kullen PJ, Frias JM (2019) Predicting quality attributes of strawberry packed under modified atmosphere throughout the cold chain. Food Packag Shelf Life 21:100354

Kamilaris A, Fonts A, Prenafeta-Boldú FX (2019) The rise of blockchain technology in agriculture and food supply chains. Trends Food Sci Technol 91:640–652

Kouhizadeh M, Saberi S, Sarkis J (2021) Blockchain technology and the sustainable supply chain: theoretically exploring adoption barriers. Int J Product Econ 231:107831

Levinson DR (2009) Traceability in the food supply chain. Retrieved from http://oig.hhs.gov/oei/reports/oei-02-06-00210.pdf. Accessed 25 Apr 2020

Lloyd J, Cheyne J (2017) The origins of the vaccine cold chain and a glimpse of the future. Vaccine 35:2115–2120

Martin G, Ronan G (2000) Managing the cold chain for quality and safety. Flair-flow Europe technical manual 378A/00. https://seafood.oregonstate.edu/sites/agscid7/files/snic/managing-the-cold-chain-for-quality-and-safety.pdf. Accessed 25 Apr 2019

Mattoli V, Mazzolai B, Mondini A, Zampolli S, Dario P (2010) Flexible tag datalogger for food logistics. Sens Actuat A: Phys 162(2):316–323

McKinnon AC, Campbell J (1998) Quick response in the frozen food supply chain: the manufacturers' perspective. Christian Salvesen logistics research paper no. 2. http://www2.hw.ac.uk/sml/downloads/logisticsresearchcentre/cs2.pdf. Accessed 25 Apr 2021

Mercier S, Villeneuve S, Mondor M, Uysal I (2017) Time-temperature management along the food cold chain: a review of recent developments. J Food Sci 16:647–667

Mirabelli G, Solina V (2019) Blockchain and agricultural supply chains traceability: research trends and future challenges. Procedia Manuf 42:414–421

Mitsugi J et al (2007) Architecture development for sensor integration in the EPC global network, AutoID Labs white paper WP-SWNET-018

MOH Singapore (2005) Guideline on how to maintain the vaccine cold chain. https://elis.moh.gov.sg/elis/info. Accessed 20 May 2021

Montanari R (2008) Cold chain tracking: a managerial perspective. Trends Food Sci Technol 19(8):425–431

Morris C, Young C (2000) "Seed to shelf", "teat to table", "barley to beer" and "womb to tomb": discourses of food quality and quality assurance schemes in the UK. J Rural Stud 16(1):103–115

NHB (2010). Technical standards and protocol for the cold chain in India, Cold storage for fresh horticulture produce requiring pre-cooling before storage (technical standards number NHB-CS-Type 02-2010), National Horticulture Board, Govt. of India

NZFSA (2003) An introduction to HACCP. New Zealand Food Safety Authority. Retrieved from https://www.mpi.govt.nz/dmsdocument/21446-An-introduction-to-HACCP-Food-Safety-Information-for-New-Zealand-Businesses. Accessed 20 Mar 2022

Olsson A (2004) Temperature controlled supply chains call for improved knowledge and shared responsibility. In: Aronsson H (ed) Conference proceedings NOFOMA 2004, Linköping, pp 569–582

Onwude DI, Chen G, Eke-emezie N, Kabutey A, Klhaled AY, Sturm B (2020) Recent advances in reducing food losses in the supply chain of fresh agricultural produce. Processes, MDPI, vol 8, p 1431. https://doi.org/10.3390/pr8111431

Opara UL (2003) Traceability in agriculture and food supply chain: a review of basic concepts, technological implications, and future prospects. Food Agric Environ 1(1):101–106

Opara UL, Mditshwa A (2013) A review on the role of packaging in securing food system: adding value to food products and reducing losses and waste. Afr J Agric Res 8(22):2612–2630

Óskarsdóttir K, Oddsson GV (2019) Towards a decision support framework for technologies used in cold supply chain traceability. J Food Eng 240:153–159

Peri C (2006) The universe of food quality. Food Qual Prefer 17(1–2):3–8

Pinto D, Castro I, Vicente A (2006) The use of TIC's as a managing tool for traceability in the food industry. Food Res Int 39(7):772–781

Rijswijk WV, Frewer LJ (2006) How consumers link traceability to food quality and safety: an international investigation. In: 98th EAAE seminar "marketing dynamics within the global trading system: new perspectives", Greece, pp 1–7

Rohr A, Luddecke K, Drusch S, Muller M, Alvensleben R (2005) Food quality and safety-consumer perception and public health concern. Food Control 16(8):649–655

Rong A, Akkerman R, Grunow M (2009) An optimization approach for managing fresh food quality throughout the supply chain. Int J Product Econ 131:421–429

Ruan J, Shi Y (2016) Monitoring and assessing fruit freshness in IOT-based e-commerce delivery using scenario analysis and interval number approaches. Inf Sci 373:557–570

Ruiz-Garcia L, Lunadei L (2010) Monitoring cold chain logistics by means of RFID. In: Turcu C (ed) Sustainable radio frequency identification solutions, Croatia, Intech, pp 37–50

Ruiz-Garcia L, Barreiro P, Rodriguez-Bermejo J, Robla JI (2007) Monitoring intermodal refrigerated fruit transport using sensor networks: a review. Spanish J Agric Res 5(2):132–156

Ruiz-Garcia L, Barreiro P, Robli JI (2008) Performance of zigbee-based wireless sensor nodes performance of zigbee-based wireless sensor nodes. J Food Eng 87:405–415

SARDI (2006) Maintaining the cold chain: air freight of perishables. South Australian Research and Development Institute

Sarig Y (2003) Traceability of food products. Agric Eng Int CIGR J Sci Res Dev. Invited overview paper

Scharff RL (2010) Health-related costs from foodborne illness in the United States, pp 1–28. Retrieved from http://www.publichealth.lacounty.gov/eh/docs/ReportPublication/HlthRelatedCostsFromFoodborneIllinessUS.pdf. Accessed 25 Apr 2019

Shan Q, Lui Y, Prosser G, Brown D (2004) Wireless intelligent sensor networks for refrigerated vehicle. In: Proceedings of IEEE 6th CAS symposium on emerging technologies: mobile and wireless communication, Shanghai, China, vol 2, pp 525–528

Shan Q, Lui Y, Prosser G, Brown D (2005) Wireless monitoring system for vehicle refrigerator. In: Proceedings of the international conference on information acquisition, June, Hong Kong and Macau. China, pp 417–420

Song JM, Sung JW, Park TH (2019) Applications of blockchain to improve supply chain traceability. Procedia Comput Sci 162:119–122

Tsang YP, Choy KL, Wu CH, Ho GTS, Lam HY (2019) Blockchain-driven IoT for food traceability with an integrated consensus mechanism. IEEE Access 7:129000–129017

Todd S (2008) Refrigerated medicinal products: what pharmacists need to know. Pharmaceut J 281:449–453

Trienekens JH, Wognum PM, Beulens AJM, van der Vorst JGAJ (2012) Transparency in complex dynamic food supply chains. Adv Eng Inform 26(1):55–65

UN (2007) Safety and quality of fresh fruit and vegetables, a training manual for trainers. United Nations

Van Reeuwijk LP (1998) Guidelines for quality management in soil and plant laboratories. FAO, Rome Publication #M-90. Retrieved from http://www.fao.org/docrep/w7295e/w7295e00.HTM. Accessed 25 Apr 2019

Vogt DU (2005) Food safety issues in the 109th congress. CRS report for congress, US

Wang SG, Wang RZ (2005) Recent developments of refrigeration technology in fishing vessels. Renew Energy 30:589–600

Wang L, Kwok SK, Ip WH (2010) A radio frequency identification and sensor-based system for the transportation of food. J Food Eng 101(1):120–129

White J (2007) How cold was it? Know the whole story. Frozen Food Age 56(3):38–40

WHO (2002) WHO global strategy for food safety: safer food for better health. Retrieved from https://apps.who.int/iris/handle/10665/42559. Accessed 25 Apr 2020

WHO (2007) Food safety and foodborne illness. World Health Organization. Retrieved from https://foodhygiene2010.files.wordpress.com/2010/06/who-food_safety_fact-sheet.pdf. Accessed 15 Apr 2019

WHO (2015a) How to monitor temperatures in the vaccine supply chain. WHO vaccine management handbook module VMH-E2. http://www.who.int/immunization_delivery/systems_policy/evm/en/index.html. Accessed 20 May 2021

WHO (2015b) Introducing solar-powered vaccine refrigerator and freezer systems: a guide for managers in national immunization programmes. ISBN 978-92-4-150986-2

WHO (2020) Immunization agenda 2030: a global strategy to leave no one behind

WRAP (2008) The food we waste. The Waste & Resources Action Programme (WRAP) report. Retrieved from http://wrap.s3.amazonaws.com/the-food-we-waste.pdf. Accessed 25 Apr 2020

Wu W, Cronjé P, Verboven P, Defraeye T (2019) Unveiling how ventilated packaging design and cold chain scenarios affect the cooling kinetics and fruit quality for each single citrus fruit in an entire pallet. Food Packag Shelf Life 21:100369

Chapter 4
Cold Chain Management Essentials

As mentioned by Casper (2007) and Zhang (2007), cold chain logistics (CCL) is comprised of equipment and processes that keep perishable products under controlled cold environment from production to consumer end in a safe, wholesome, and good-quality state. Typically, cold chain needs to deal with refrigerated facilities such as refrigerators, cold storage warehouses, refrigerated truck and containers in the stage of production, processing, packaging, warehousing, transportation, distribution, and retailing.

However, cold chain is not only about the facilities and equipment used in different stages. The integrity of the cold chain must be maintained from the very beginning of production or processing, through each link such as loading, unloading, transport, handling, and storage to the consumer end (Salin and Nayga 2003). To manage an integrated cold chain, the basic understanding of refrigeration and freezing, product characteristics, necessary facilities, and equipment is necessary.

4.1 Basic Elements for Cold Chain

4.1.1 Refrigeration and Freezing

Food preservation is one of the oldest technologies used by human beings. The different preservation techniques are commonly used today, including refrigeration and freezing, canning, irradiation, dehydration, salting, pickling, pasteurizing, fermentation, carbonation, and chemical preservation, etc. Among them, refrigeration and freezing are useful and widespread means of preserving the quality of foodstuffs and thus protecting consumer health, but require accurate control of the cold chain, from the producer to the consumer (Coulomb 2008).

© Springer Nature Switzerland AG 2023
M. M. Aung and Y. S. Chang, *Cold Chain Management*, Springer Series in Advanced Manufacturing, https://doi.org/10.1007/978-3-031-09567-2_4

Refrigeration plays an essential role in our daily life and offers benefits in a huge range of fields, particularly in the food, health, and indoor environment fields (IIR 2007). In addition, it is commonly used in chemical and process industries to separate gases and solutions, to remove the heat of reaction, and to control pressure by maintaining low temperature. It is also used commonly in the beverage industry, in medicine, and even in the storage of furs and garments. Furs and wool products are commonly stored at 1–4 °C to protect them against insect damage. Refrigeration maintains quality and prolong shelf life by keeping the product temperature at the point where metabolic and microbial deteriorations are minimized. Microorganisms such as bacteria, yeasts, molds, and viruses cause off-flavors and odors, slime production, changes in the texture and appearance, and eventual spoilage of foods. Refrigeration slows down the chemical and biological processes in foods and the accompanying deterioration and the loss of quality. The storage life of fresh perishable foods such as meats, fish, fruits, and vegetables can be extended by several days by cooling and by several weeks or months by freezing (Stoecker 1998).

Refrigeration plays a vital role in reducing postharvest losses by slowing down bacterial growth, increasing shelf life, and preserving the nutritional and organoleptic properties of foodstuffs. No other processing technology combines the ability to extend product shelf life and in parallel maintain the initial physical, chemical, nutritional, and sensory properties desired by consumers to the same extent as refrigeration. Over 13% of food in the world, an important amount of food losses occurs in global food supply chain mainly due to a lack of refrigeration. A more efficient cold chain can significantly reduce food losses and thus improve food safety and security in a sustainable way. Greater use of refrigeration technologies would ensure better worldwide nutrition, in terms of both quantity and quality. An improvement of the cold chain would save over 475 million tons of food that could theoretically feed 950 million inhabitants per year. International Institute of Refrigeration (IIR) recommends that refrigeration is the best technology, with no associated risks, to ensure the safety of foods through chilling or freezing. However, compliance needs to be ensured at all stages in the cold chain, without which foodborne diseases and series of deaths occur, even in the most developed countries (IIR 2008, 2009, 2020).

Freezing is one of the oldest and most widely used methods of food preservation, which allows preservation of taste, texture, and nutritional value in foods better than any other method. The freezing process is a combination of the beneficial effects of low temperatures at which microorganisms cannot grow (i.e., stop bacterial action), chemical reactions are reduced, and cellular metabolic reactions are delayed (Delgado and Sun 2000). Freezing has been successfully employed for the long-term preservation of many foods, providing a significantly extended shelf life. The process involves lowering the product temperature generally to −18 °C or below (Fennema et al. 1973). The physical state of food material is changed when energy is removed by cooling below freezing temperature. The extreme cold simply retards the growth of microorganisms and slows down the chemical changes that affect quality or cause food to spoil (George 1993).

Table 4.1 highlights that the number of refrigeration, air conditioning, and heat pump system in operation worldwide is roughly 5 billion, including 2.6 billion air-conditioning units (stationary and mobile), and 2 billion domestic refrigerators and freezers. Global annual sales of refrigeration, air conditioning, and heat pump equipment amount to roughly 500 billion USD, this being roughly about three quarters of global supermarket sales. The refrigeration sector (including air conditioning) consumes about 20% of the overall electricity used worldwide.

Table 4.1 Number of refrigeration systems in operation worldwide per application (IIR 2019) (Reprinted by permission of IIR)

Applications	Sectors	Equipment	Number of units in operations
Refrigeration and food	Domestic refrigeration	Refrigerators and freezers	2 billion
	Commercial refrigeration	Commercial refrigeration equipment (including condensing units, stand-alone equipment, and supermarket systems)	120 million
	Refrigerated transport	Refrigerated vehicles (vans, trucks, semi-trailers, or trailers)	5 million
		Refrigerated containers ("reefers")	1.2 million
	Refrigerated storage	Cold stores	50,000
Air conditioning	Stationary air conditioning	Residential air-conditioning units	1.1 billion
		Commercial air-conditioning units	0.5 billion
		Water chillers	40 million
	Mobile air-conditioning systems	Air-conditioned vehicles (passenger cars, commercial vehicles and buses)	1 billion
Refrigeration and health	Medicine	Magnetic resonance imaging (MRI) machines	50,000
Refrigeration in industry	Liquefied natural gas (LNG)	LNG regasification terminals	126
		LNG tanker fleet (vessels)	525
Heat pumps		Heat pumps (residential, commercial, and industrial equipment, including reversible air-to-air conditioners)	220 million
Leisure and sports		Ice rinks	17,000

The data and figures collected in different years show that the use of refrigeration is increasing regularly and will no doubt continue to increase in the future (IIR 2007, 2015, 2019). IIR stated that the role of refrigeration in sustainable development may be appreciated through its social, economic, and environmental dimensions.

Social Dimension

- the refrigeration sector generates jobs (12 million people are employed worldwide in the refrigeration sector);
- refrigeration is indispensable to human life (e.g., in health care and food);
- air conditioning contributes to social development [in the US, employment of mechanics and installers in heating, refrigeration, and air conditioning is projected to grow by 21% from 2012 to 2022, much faster than the average for all occupations (11%)].

Economic Dimension

- refrigeration is necessary for the implementation of many current or future energy sources;
- many industrial processes could not operate without refrigeration;
- many cutting-edge technologies require refrigeration technologies;
- refrigeration is vital to reduce postharvest and postslaughtering losses and in the preservation of food products;
- the lack of cold chain causes significant global food losses: up to almost 20% of the global food supply.

Environmental Dimension

- in maintaining biodiversity by cryopreservation of genetic resources;
- refrigeration technology provides basic for heat pumps to save energy and carbon emissions in all kinds of industrial and building applications;
- in enabling the liquefaction of carbon dioxide (CO_2) for underground storage.

On the other side, atmospheric emissions of certain refrigerant gases used in installations and energy consumption that contribute to CO_2 emission are adverse environmental effects of refrigeration. However, with the advancement of technology, refrigeration and heat pumps are among the environment-friendly technologies that may use renewable energy.

4.1.2   Insight into Refrigeration and Freezing of Perishable Foods

Refrigeration is crucial for the food sector because it ensures optimal preservation of perishable foodstuffs and provides consumers with safe and wholesome products. Global food production comprises roughly one-third of perishable products requiring refrigeration. In 2010, out of a total global food production (agricultural commodities, fish, meat products, and dairy products) of 6300 million tons, only about 400 million tons were preserved using refrigeration (in chilled and frozen state), while about 2000 million tons required refrigerated processing. Continuous and ubiquitous refrigeration is necessary throughout the perishable food chain to reduce the loss in quality and value of food products (IIR 2015).

Perishable food products are transported by trucks and trailers, railroad cars, ships, airplanes, or a combination of them from production areas to distant markets. Transporting large quantities over long distances usually requires strict climate control by refrigeration and adequate ventilation and adequate insulation to keep the heat transfer rates at reasonable levels. Several methods of refrigeration are used in transportation of CCL such as ice, ice and salt, dry ice, hold-over plate systems, cryogenic systems, and mechanical refrigeration. Today, among them, mechanical refrigeration is the predominant type. Trailers refrigerated with cryogenic refrigerants, usually liquid CO_2 or nitrogen (N_2), have been used to some extent, but they are not as popular trailers refrigerated mechanically (Brecht et al. 2019).

Refrigeration removes excess heat and provides temperature control to keep food products cool in transport vehicles. Heat always radiates or flows toward the cold or refrigeration source. In the US, heat is measured in British thermal units (Btu). The Btu is defined as the amount of heat required to raise the temperature of 1 lb. (0.45 kg) of water at 1 °F (0.56 °C). A metric equivalent of the Btu is the kilojoule (kj) or 1 Btu = 1.005 kJ.

Heat is a measurable form of energy that flows from a high-temperature source to a low-temperature source. The heat transfer can occur by three means: conduction, convection, and radiation. **Conduction** is the movement of heat through solid objects, such as the transit vehicle's insulated walls, ceilings, and doors, or between solid objects that are in direct contact with each other, such as from cargo to the walls of the cartons containing the cargo. **Convection** occurs when warmer areas of gases and liquids rise to cooler areas of the gas or liquid, creating convection currents that mix the warmer and cooler areas. For instance, refrigerated airflow moves heat generated from respiring produce and heat transferred from interior walls to the transport vehicle's cooling unit. **Radiation** is the transfer of heat by electromagnetic waves that can travel through a gaseous medium or even a vacuum, such as the transfer of radiant heat generated from the sun or hot roadways to the exterior of the transport vehicle.

James and James (2014) added that **evaporation/condensation** is another mean of heat transfer occurred when a liquid is changed to a vapor. For most foods, the heat loss can happen through evaporation of water from the surface, which is a minor component of the total heat loss, though it is the major component in vacuum cooling. The above four modes of heat transfer can occur individually or combined. In many applications, all four occur simultaneously but with different levels of importance.

Three types of heat are important in refrigerated transport of perishable cargo. These include:

1. sensible heat;
2. latent heat;
3. respiratory heat (only for fresh horticultural items).

Sensible heat is heat exchanged by a thermodynamic system or body that changes the temperature of the system or body. As the name implies, sensible heat is the heat that you can feel. Latent heat is the heat required to convert a solid into a liquid or vapor, or a liquid into a vapor, without a change of temperature. Stated differently, latent heat is the energy absorbed or released by a thermodynamic system during a constant temperature process. Examples include ice melting or water boiling. Sensible heat can be felt, while latent heat is the type of heat that cannot be felt.

Respiratory heat is the heat generated in fresh produce primarily developed through respiration, a process where fresh fruits and vegetables use energy from stored reserves and oxygen from the surrounding air to keep alive, even after harvest. Heat, known specifically as "vital heat", is released as a by-product of the respiration process (Kader 2002), which contributes to the refrigeration needs that must be considered in designing storage rooms and during transportation of fresh produce (Saltveit 2016). The information about respiration process is explained further in Sects. 4.2.5 and 7.2: respiratory metabolism.

Refrigeration Methods

– Cooling

The nature of a Transport Refrigeration Unit, commonly called a "reefer unit", is that a temperature differential exists between the air entering the reefer unit, (i.e., the return air), and the air exiting the unit, (i.e., the discharge or supply air) due to heat removal or addition as the return air passes through the reefer unit. The cargo itself and the air surrounding the cargo are exposed to a range of temperatures and atmospheres during transit. Controlling, modifying, and monitoring the characteristics of environments inside insulated boxes are critical to maximizing the shelf life of the cargo and, in the case of perishable foods, to optimizing the wholesomeness, safety, and quality of the food.

Mechanical refrigeration operates by absorbing heat at one point and dispensing it at another. The refrigerant picks up heat through a coil (evaporator) inside the cargo space and discharges it through another coil (condenser) on the outside. The refrigerant is circulated through the system by a compressor. Transport reefer units are equipped with microprocessors that are interfaced with temperature sensors controlling the supply air and return air temperatures that are located inside the reefer unit (Fig. 4.1).

Refrigerants are chosen based on their high heat capacity and ability to change phase between liquid and gas at the desired temperature. The refrigeration capacity needed for a particular load depends on the desired product temperature, the amount

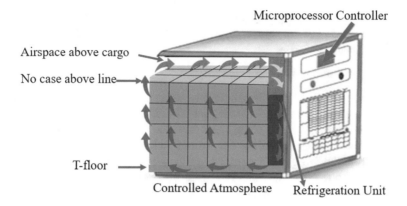

Fig. 4.1 Air circulation in the reefer box

of vehicle insulation, ambient temperatures, product temperature at loading, amount of product respiration heat, and the extra capacity (reserve) desired.

– **Defrost**

A refrigeration unit's evaporator coil facilitates heat transfer from the circulating air from the cargo compartment to a refrigerant that is circulated within the refrigeration system. When moisture carried from warmer air returns from the cargo compartment, it condenses on the cooler evaporator coil surface, causing moisture or ice to accumulate on the coil. Ice accumulation on the evaporator coil will eventually adversely impact conditioned airflow and refrigeration capacity. To correct this condition, manual, periodic (timed), and "on demand" defrosts are required to clear the ice from the evaporator coil. Failure to adequately defrost the coil can result in poor temperature management, quality and food safety issues, and load rejections.

– **Ice**

Crushed or slush ice blown over the top of produce loads is used to refrigerate and maintain high levels of humidity for certain produce items. This is known as "top icing". Some shippers may apply the crushed ice or slush ice to individual pallet loads or in individual boxes of product (package icing) before loading. For package icing, up to 10 lbs (4.5 kg) of crushed ice may be placed in the cartons after packing. Ice is sometimes used to cool fruits and vegetables in the field and during short transit.

– **Cryogenic Refrigerants**

Cryogenic refrigerating systems, which use liquid or solid CO_2 or liquid N_2, have been available for highway trailers for many years. They are utilized primarily in delivery operations requiring 1 day or less transit time, since supplies of liquid cryogens are generally not available at truck stops. The benefits of cryogenic transport are that they have fewer moving parts to maintain and replace and also allow quick recovery of thermostat set-point temperature after delivery stops. Cryogenic refrigeration systems have been mostly replaced by multi-temperature refrigeration systems for short-distance deliveries.

Refrigeration Load of Cold Storage Rooms

Refrigerated spaces are maintained below the temperature of their surroundings, and thus, there is always a driving force for heat flow toward the refrigerated space from the surroundings. As a result of this heat flow, the temperature of the refrigerated space will rise to the surrounding temperature unless the heat gained is promptly removed from the refrigerated space. A refrigeration system should obviously be large enough to remove the entire heat gain in order to maintain the refrigerated space at the desired low temperature. Therefore, the size of a refrigeration system for a specified refrigerated space is determined based on the rate of heat gain of the refrigerated space.

The total rate of heat gain of a refrigerated space through all mechanisms under peak (or times of highest demand) conditions is called the refrigeration load, and it consists of

(1)   transmission load, which is heat conducted into the refrigerated space through its walls, floor, and ceiling;
(2)   infiltration load, which is due to surrounding warm air entering the refrigerated space through the cracks and open doors;
(3)   product load, which is the heat removed from the food products as they are cooled to refrigeration temperature;
(4)   internal load, which is heat generated by the lights, electric motors, and people in the refrigerated space;
(5)   refrigeration equipment load, which is the heat generated by the refrigeration equipment as it performs certain tasks such as reheating and defrosting.

The size of the refrigeration equipment must be based on peak refrigeration load, which usually occurs when the outside temperature is high and the maximum amount of products is brought into the cool storage room at field temperatures. The details of calculation on refrigeration load can be found in Stoecker (1998).

Some fresh fruits, vegetables, and carcass meats are shipped before they have been precooled to the proper transit temperature. The ideal situation would be for the carrier to accept only properly precooled product. When this is not practical, the trailer's refrigeration system must bear the additional heat load. Truck refrigeration units may have enough reserve capacity to remove a reasonable amount of heat in addition to respiration heat and heat transferring through the vehicle body. However, if the product is much above the desired transit temperature at loading time, the entire heat load should be estimated. If the estimated heat load is more than the refrigeration unit is rated to bear, the trailer should not be used. The sample calculations of cold store refrigeration load can be found in Jonston et al. (1994), Brecht et al. (2019), and Yuzainee et al. (2019).

Refrigeration and Freezing of Foods

Stoecker (1998) mentioned that the refrigeration and freezing processes for perishable food are complex as there are specific handling processes for different commodities. The ordinary refrigeration of foods involves cooling only without any phase change. The freezing of foods, on the other hand, involves three stages: cooling to the freezing point (removing the sensible heat), freezing (removing the latent heat), and further cooling to the desired subfreezing temperature (removing the sensible heat of frozen food), as shown in Fig. 4.2.

The rate of freezing has a major effect on the size of ice crystals formed and the quality, texture, and nutritional and sensory properties of many foods. During slow freezing, ice crystals can grow to a large size, whereas during fast freezing, a number of much smaller ice crystals start forming at once. Large ice crystals are not desirable since they can puncture the walls the cells, causing a degradation of texture and a loss of natural juices during thawing.

Some fruits and vegetables experience undesirable physiological changes when exposed to low (but still above-freezing) temperatures, usually between 0 and 10 °C. The resulting tissue damage is called the **chilling injury** that exhibits internal discoloration, soft scald, skin blemishes, soggy breakdown, and failure to open. It differs from **freezing injury**, which is caused by prolonged exposure of the fruits and vegetables to subfreezing temperatures and thus the actual freezing at the affected areas. Products near the refrigerator coils or at the bottom layers of refrigerator cars and

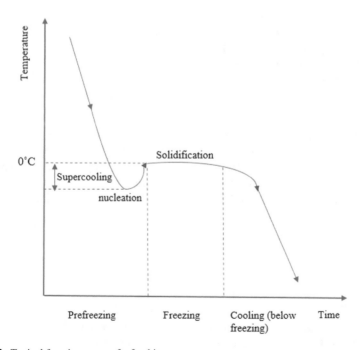

Fig. 4.2 Typical freezing curve of a food item

trucks are most susceptible to freezing injury, and it can be solved by heating cars and trucks and providing adequate air circulation in cold storage rooms.

The thermal properties of foods are dominated by their water content. Fresh fruits and vegetables are live products, and they continue to respire after harvesting. Heat of respiration is released during this exothermic reaction. **Dehydration** or moisture loss causes a product to shrivel or wrinkle and lose quality. The loss of moisture from fresh fruits and vegetables is also called **transpiration**. The transpiration rate is different for various kinds of fruits and vegetables. The transpiration rate varies with the environmental conditions such as the temperature, relative humidity, and air motion. This process can be minimized by keeping low temperature, high relative humidity, and appropriate level of air circulation. One of the methods used to reduce is **waxing**, but thick wax coating may increase decay, especially when no fungicides are used.

Precooling Methods

The primary purpose of refrigeration and freezing for perishable goods is to keep the goods under proper low temperature, slowing down the deterioration and loss of quality. The storage life of fresh perishable foods can be extended by several days by refrigeration (cooling) and several weeks or months by freezing. In the white paper published by Postharvest Education Foundation (PEF), Kitinoja (2013) summarized cooling technologies and where they can be applied. Cooling provides the following benefits for perishable horticultural foods:

- reduces respiration: lessens perishability;
- reduces transpiration: lessens water loss, less shriveling;
- reduces ethylene production: slows ripening;
- increases resistance to ethylene action;
- decreases activity of microorganisms;
- reduces browning and loss of texture, flavor, and nutrients;
- delays ripening and natural senescence.

Precooling is necessary to make the food cool for handling, processing, storage, and transport as it removes the field heat from the product after harvest. Operators can choose from simple farm-based methods such as **using ice**, to more complex systems like **forced air, hydrocooling, or vacuum cooling**.

According to Mercier et al. (2017), there exists a variety of precooling techniques, including

- forced-air cooling;
- hydrocooling;
- room cooling;
- vacuum cooling;
- cryogenic cooling.

Forced-air cooling is usually accomplished by creating an air pressure differential on two sides of a stacked pallet tunnel. The air pressure allows the cooled air to circulate from the outside through the warmer product and back into the room after going through the refrigeration unit for recooling. Hydrocooling uses chilled or cold water to lower the temperature of the product in pallets or large containers prior to further packaging. Room cooling is a relatively older technique in which the product is placed in either boxes or pallet inside a cold room and exposed to cooler temperatures. Room cooling takes longer and is sometimes used after other precooling techniques for temperature stabilization. Vacuum cooling is used in situations where rapid cooling of the product is required or desired; significant temperature drops are achieved in a short time frame by evaporating the moisture from the product. In vacuum cooling, the pressure is dropped to the point where water can boil at a very low temperature, in order to encourage faster evaporation. Cryogenic cooling is accomplished by using liquid N_2 or dry ice, which produces boiling temperatures as low as -196 °C. As the product moves along a conveyor belt through a tunnel where liquid N_2 is evaporating, the product's temperature drops rapidly. If this conveyor tunnel is designed to be vertical, it is called a spiral chiller and is generally used for frozen products.

There are considerations concerning precooling products and equipment (Brecht et al. 2019):

- The perishable product must be precooled to the shipper's specified temperature prior to loading. Transit vehicles are not designed to effectively precool chill or frozen cargos!
- If the interior of the transit vehicle is hot, the cargo can potentially be temperature abused by contact with hot sidewalls and floors of the containers.
- Precooling the transit vehicle (with the doors closed) can help suppress temperature-induced damage to the perishable cargos from hot floors and sidewalls.
- Loading transit vehicles in a hot humid and open environment is a bad practice. Refrigerated loading docks with cold tunnels are recommended. Cold tunnels prevent outside ambient air from entering the refrigerated dock and the interior of precooled transit vehicles.
- In open conditions with hot humid air:

  - Condensation can form on the exposed cartons ("cargo sweat") when the supplier moves the refrigerated cargo from the cold room to a hot, humid dock, or open space. Under hot, humid conditions, the problem of cargo sweat will likely persist even if the transit vehicle is not precooled.
  - Condensation can occur near the doors of the precooled transit vehicle and on the ceiling. The concern is that moisture on the ceiling might fall on the cartons.

The quality of precooling is strongly dependent on the packaging. Both the design and the material of the packaging need to be considered when choosing the optimal cooling method. The design of the package vent areas directly affects the flow of cold fluid inside the pallet as well as the heat and mass transfer rates at the surface of the food. The package should be designed not only for sufficient protection against mechanical damage along the cold chain but also for optimal precooling, otherwise precooling uniformity and efficiency will be decreased.

For storage, there are options for food handlers that range from small walk-in cold rooms to large-scale commercial refrigerated warehouses. Food processors can choose from chillers, blast freezing, individual quick freezing (IQF), freeze drying, and many other technologies. During transport, cold can be provided via the use of ice, trailer-mounted refrigeration systems, evaporative coolers, or via passive cooling technologies (insulated packages or pallets covers during transport).

The suitability of these options will depend upon the food products being handled and the level of sophistication of the value chain. The factors might include product itself (e.g., product's physical or chemical properties and its sensitivity to chilling or freezing injuries) or other factors such as harvest volume, higher customer satisfaction, and economic consideration. In general, the highest cost will be for mechanical refrigeration systems using electricity or diesel fuel where temperatures are the hottest, but the benefits of using cold chain technologies can still outweigh costs, since it is in these regions where food losses due to lack of temperature management are the highest (Mercier et al. 2017).

Freezing Methods

Common freezing methods are summarized by Stoecker (1998), and it includes **air-blast freezing**, where high-velocity air at about -30 °C is blown over the food products; **contact freezing**, where packaged or unpackaged food is placed on or between cold metal plates and cooled by conduction; **immersion freezing**, where the food is immersed in low-temperature brine or another fluid; and **cryogenic freezing**, where food is placed in a medium cooled by a cryogenic fluid such as liquid N_2 or solid CO_2; and the combination of the methods above.

First, air-blast freezing method is in most common use by refrigerated warehouses for freezing foods—either from the unfrozen state for a processor with limited freezer capacity or for bringing the temperature of still-frozen foods back to -18 °C after they have been exposed to higher than optimal temperatures. The freezing process can be speeded up even further by using a free-flow freezing process to achieve IQF product pieces. The unpackaged food is frozen either on belt freezers where air at -40 °C blows up through a mesh belt and through a thin layer of small food product pieces or in fluidized-bed freezers. Second, contact freezing is limited to flat foods not thicker than about 8 cm with good thermal conductivity, such as meat patties, fish filets, and chopped leafy vegetables.

Third, immersion and cryogenic freezing methods are quick freezing methods using liquid N_2; however, the technologies are extremely expensive. Methods that produce quick freezing (IQF, liquid N_2) result in better-quality food products than the methods that provide slow freezing (traditional freezer room racking). In summary, rapid freezing prevents undesirable large ice crystals forming in the frozen food product because the molecules do not have time to form. Slow freezing creates large, disruptive ice crystals. During thawing, they damage the cells and break cell walls and membranes. This causes vegetables to have a mushy texture and meats to weep and lose juiciness. Quicker freezing methods have advantages, however, also can be more expensive (Kitinoja 2013).

Cost Analysis of a Freezing Application

An accurate cost analysis of a freezing application can only be done on a case-by-case basis. Johnston et al. (1994) presented some useful data for freezing cost analysis, and many factors have to be considered that depend on local conditions and economics. Costs may include, but are not limited to, the following:

- initial cost of freezer;
- lost interest in cash outlay of freezer;
- maintenance and repair;
- electricity needed to reach and maintain freezing temperatures;
- packaging materials;
- water and fuel to prepare food for freezing;
- added ingredients, such as sugar or anti-darkening agents.

According to Barbosa-Cánovas (2005), the total cost of a freezing plant is generally divided into two main areas: *investment cost and production cost*. Investment cost includes *general preoperation* and *asset* expense, while production cost consists of *variable* and *fixed* costs.

For further information about the cost analysis of freezing, the details are explained in published Food and Agriculture Organisation of the United Nations (FAO) papers: Johnston et al. (1994) and Barbosa-Cánovas et al. (2005).

4.2 Product Characteristics

All cold chain products are not the same. Each product has its own unique requirements in pre- and postproduction stages such as harvesting, slaughtering, packaging, storage, transportation, and retail and could be impacted differently by external conditions.

The variety of the products and their diversified requirement in temperature or humidity contribute to the complexity of cold chain management (Ames 2006). For example, in food industry, temperature requirements vary among food items, whether frozen or chilled, and they even differ across types of frozen foods. Ice cream must be held at a lower temperature than frozen vegetables. Before setting up a cold chain system, it is necessary for logistics managers to know the product characteristic in the cold chain. Ideal storage preserves as much of the freshness as possible (Crisosto et al. 1995).

4.2.1   Temperature/Humidity

All kind of products have their own characteristics and are sensitive to storage conditions. Temperature is the key to keep the quality and the integrity of product after production/harvest in cold chain. Temperature requirement varies across types of products. In the food industry, they can be identified as frozen, cold chill, medium, and exotic chill (Smith and Sparks 2004):

- frozen is −25 °C for ice cream, −18 °C for other foods and food ingredients;
- cold chill is 0 to +1 °C for fresh meat and poultry, most dairy and meat-based provisions, most vegetables and some fruit;
- medium chill is +5 °C for some pastry-based products, butters, fats, and cheeses;
- exotic chill is +10 to +15 °C for potatoes, eggs, exotic fruit, and bananas.

Another classification proposed by logistics service providers is found to have five major categories: hot food (above 60 °C constantly), fresh food (18 °C constantly), cold food (0 to +7 °C), chilled food (−2 to +2 °C), frozen food (below −18 °C), and deeply frozen food (below −30 °C) (Kuo and Chen 2010).

If a food supply chain is dedicated to a narrow range of products, the temperature will be set at the level for that product set. However, many factors may influence this, if there are not sufficient refrigerated facilities in the cold chain, an optimum or compromised temperature or limited number of temperature settings is used. Failure to keep products in an appropriate temperature regime throughout their life will shorten the life of the products or adversely affect their quality or fitness for consumption.

Actually, not only temperature but also other environmental parameters like relative humidity, pressure, and light are essential to consider as some products are sensitive with them to keep their quality and safety. Low relative humidity may dry up the product, or high relative humidity may increase the water activity growing molds.

According to Thompson and Kader (2001), fresh fruits and vegetables are categorized into three groups regarding appropriate storage requirements (Table 4.2):

- 32–36 °F/0–2 °C 90–98% humidity;
- 45–50 °F/7–10 °C 85–95% humidity;
- 60–65 °F/16–18 °C 85–95% humidity or with no humidity requirement.

Table 4.2 mentioned that humidity segments go with the temperature segments. Attention must be paid to odor producer/sensitive and ethylene producer/sensitive products.

Table 4.2 Compatible fresh fruits and vegetables during 7 days storage (Thomson and Kader 1999)

	Group 1A and 1B 0–2 °C, 1A: 90–98% rh, 1B: 85–95% rh			Group 2 7–10 °C and 85–95% rh	Group 3 13–18 °C and 85–95% rh
Vegetables	Alfalfa sprouts	Chinese cabbage*	Mint* 1A	Basil*	Bitter meleon
	Amaranth*	Chinese turnip	Mushroom	Beans, snap, green, wax	Boniato*
	Anise*	Collard*	Mustard greens*	Cactus leaves (nopales)*	Cassava
	Artichoke	Corn; sweet, baby	Parsley*	Calabaza	Dry onion
	Arugula*	Cut vegetables	Parsnip	Chayote*	Ginger
	Asparagus*	Daikon*	Radicchio	Cowpea (southern pea)*	Jicama
	Beans: fava, lima	Endive*-chicory	Radish	Cucumber*	Potato
	Bean sprouts	Escarole*	Rutabaga	Eggplant	Pumpkin
	Beet	Fennel*	Rhubarb	Kiwano (horned melon)	Squash; winter (hard rind)*
	Belgian endive*	Garlic	Salsify	Long bean	Sweet potato*
	Bok choy*	Green onion*	ScorzonEra	Malanaga*	Taro (dasheen)
	Broccoli*	Herbs* (not basil)	Shallot*	Okra*	Tomato; ripe, partially ripe, and mature green
	Broccoflower*	Horseradish	Snow pea*	Pepper, bell, chili	Yam*
	Brussels sprouts*	Jerusalem artichoke	Spinach*	Squash, summer (soft rind)*	
	Cabbage*	Kailon*	Sweet pea*	Tomatillo	
	Carrot*	Kale*	Swiss chard*	Winged bean	
	Cauliflower*	Kohlrabi	Turnip		
	Celeriac	Leek*	Turnip Greens*		
	Celery*	Lettuce*	Waterchestnut		
	Chard*		Watercress*		

(continued)

Table 4.2 (continued)

	Group 1A and 1B 0–2 °C, 1A: 90–98% rh, 1B: 85–95% rh			Group 2 7–10 °C and 85–95% rh		Group 3 13–18 °C and 85–95% rh	
Fruits and melons	Apple	Elderberry	Prune* 1B	Avocado, unripe	Lime*	Atemoya	Rambutan
	Apricot	Fig	Quince*	Babaco	Limeqat	Banana	Sapodilla
	Avocado, ripe	Gooseberry	Raspberry	Catus pear, tuna	Mandarin	Breadfruit	Sapote
	Barbados cherry	Grape	Strawberry	Calamondin	Olive	Canistel	Soursop
	Blackberry	Kiwifruit*		Carambola	Orange	Casaba melon	
	Blueberry	Loganberry		Cranberry	Passion fruit	Cherimoya	
	Boysenberry	Longan		Custartd apple	Pepino	Creshaw_melon	
	Caimito	Loquat		Durian	Pineapple	Honeydew_melon	
	Cantaloupe	Lychee		Feijo	Pummelo	Jaboticaba	
	Cashew apple	Nectarine		Granadilla	Sugar apple	Jackfruit	
	Cherry	Peach		Grapefruit*	Tamarillo	Mamey	
	Coconut	Pear: (Asian and European)*		Guava	Tamarind	Mango	
	Currant	Persimmon*		Juan Canary	Tangelo	Mangosteen	
	Cut fruits	Plum		Melon	Tangerine	Papaya	
	Date	Plumcot		Kumquat	Ugli fruit	persian Melon	
	Dewberry	Pomegranate		Lemon*	Watermelon	Plantain	

Copyright © 2001 Regents of the University of California. Used by permission

Ethylene level should be kept below 1 ppm in storage area

* Sensitive to ethylene damage

4.2.2 Time

Time in the cold chain refers to the shelf life and lead time in each link of the product flow. Environmental parameters such as temperature, relative humidity, and vibration/shock are variables that will affect the shelf life.

A US benchmark survey reports that due to the shelf life (expiration date) issues, an estimated $900 million worth of inventory was wasted in cold chains in 2001(R2N 2003). It is common practice that product shelf life is assumed in terms of average distribution conditions and no temperature variations that occur through different stages of the supply chain (Sahin et al. 2007). Suppliers face the common dilemma about deciding product shelf life. Making the shelf life shorter for assuring higher service levels may lead to huge losses on overdue products, while extending the expiration date may risk an increase in perished products.

Perishable products must move as fast as possible to provide the best quality to the consumers. The commitment of keeping a satisfied consumer requires delivery at an appropriate shelf life and with the proper routing through the chain. These are crucial factors for executing and maintaining competitive advantage. Longer lead times in supply chain can lead to the result of damaging perishable goods. Items in a critical state can be assigned to shorter transport routes to prevent losses and to provide consistent quality to customers.

4.2.3 Ethylene Producers/Ethylene Sensitive

For fresh produce, temperature, humidity, and odor are three factors that need to be considered before putting the produce into storage. Ethylene compatibility is another factor to be considered because this can induce a negative effect on plants' freshness and shelf life. For example, this can result in senescence, over-ripening, and accelerated quality loss (Martinez-Romero et al. 2007).

Ethylene is a colorless natural organic compound, and ripening and diseased plant tissues are a significant source of ethylene. The presence of ethylene in the environment may have either beneficial or detrimental effects on harvested fruits, vegetables, and ornamentals. For example, ethylene is often used to accelerate the ripening in green tomatoes or bananas. In other instances, the presence of ethylene is undesirable as it can contribute to the yellowing of green vegetables and accelerate the ripening and death of certain crops (IATA 2011). Therefore, it is important to have the knowledge on incompatible fruits and vegetables or products which are not appropriate for storing in the same place.

4.2.4   Chilling and Freezing Injury

In USDA agriculture handbook no. 66, Wang and Wallace (2003) describe guidelines on commercial storage of fruits, vegetables, and florist and nursery stocks. Many fruits and vegetables of tropical or subtropical origin are sensitive to low temperatures. These crops are injured after a period of exposure to chilling temperatures below 10–15 °C (50–59 °F) but above their freezing points. Certain horticultural crops of temperate origin are also susceptible to chilling injury. Those temperate crops, in general, have lower threshold temperatures of <5 °C (41 °F). At these chilling temperatures, the tissues weaken because they are unable to carry on normal metabolic processes. Symptoms may be pitting, discoloration, off-flavors, physiological deterioration, and increased decay. Chilling injury varies with both time and temperature. Some commodities, such as bananas, will be injured by a few hours' exposure to chilling temperatures. Others can be held below the desired storage temperature for several days before incurring serious injury. Based on maximum transit time of 5 days, however, chilling injury may vary considerablly by cultivar, harvest season, holding time, maturity of commodity, etc.

Freezing injury occurs when ice crystals form in the tissues. Cultivars, locations, and growing conditions may affect the freezing point. Freezing fresh meat will darken its color and increase thaw drip. Eggs may crack and incur irreversible physical changes by freezing. The texture of some cheeses is changed by freezing. All fruits and vegetables can be categorized into three groups (most, moderately, and least susceptible) based on their sensitivity to freezing to lessen freezing injury with them. Freezing losses are most common in fruits and vegetables such as apples (moderately susceptible) and lettuce (most susceptible) which normally subject to injure at temperatures near their freezing point. The extent of injury varies with the characteristics of the product and the severity of freezing. Commodities such as beets and cabbage can withstand light freezing and thawing several times without permanent injury. Other products such as potatoes and tomatoes are permanently injured by only one slight freezing. Once the injury occurs, whether it is much or little, the products are more susceptible to decay. To be on the safe side, the highest temperature at which freezing of a specific commodity may occur should be used as a guide for recommending the optimum storage temperature.

4.2.5   Respiratory Metabolism

Fresh fruits and vegetables are alive after postharvest and carry on processes characteristic of all living things. One of the most important of these is respiratory metabolism. Commodities and cultivars with higher rates of respiration tend to have shorter storage life than those with lower rates of respiration.

Some products have ventilation requirements for storage and a specific storage life. Correct rates of ventilation are vital to ensure ethylene and CO_2 levels do not increase and speed the deterioration of the produce. Ventilation requirements mostly depend on product rate of respiration and sensitivity to ethylene and moisture loss.

The storage requirements of fresh fruits are more complicated than those of fresh vegetables. Fresh fruits have different ripeness stages with corresponding temperature requirements. Because of the characteristics of highly perishable goods, shelf life is given for some products. To better manage fruits and vegetables in cold chain, it is essential to know the perishability rate of the commodity. For example, vegetables such as broccoli are found to have very high perishability rating compared to cauliflower (high), cabbage (moderate), and onion (low). Most of the storages are not stackable, and some fruits need plenty of air circulation. Most fruits are sensitive to temperature fluctuation, ethylene, odor, and bruise easily.

While fresh fruits and vegetables are usually displayed on the open shelf at room temperature, poultry and meat are displayed on the shelf with chilled air. Frozen food such as frozen pizza and ice cream may run the risk of customers opening the freezer door frequently. However, these problems are solved by equipping freezers with a temperature alarm and having quick turnover of frozen goods in the supermarket. The high perishables business incurs high risk. Hence, the characteristics of these fruits, as described above, require a careful handling and monitoring system.

4.3 Facilities and Equipment in Cold Chain

To maintain the temperature all the way from producers/manufacturers to the consumer end, specialized facilities and technologies are essential to build a robust cold chain. Refrigerated processing, storage, and transport facilities are essential in the cold chain, to supply the consumer with safe, high-quality perishable goods.

According to Salin and Nayga (2003), highways, ports, information infrastructures, reliable electric power systems, and laws and regulations are in the upper level which may determine the whole cold chain environment in a country. These are established or enacted by government authorities. However, the development of these infrastructures should be carefully considered even though it is relevant to the development and economic power of a country. The refrigerated containers, cold storage facilities, and refrigerated vehicles are generally owned by carriers, public or private refrigerated warehouses, and trucking companies, respectively. A small part of the facilities is owned by the shippers themselves. This is decided by the outsourcing strategy.

The cold chain facility is one of the factors which differentiates a firm and allows it to compete with other companies. The adequate facilities used for a sequence of logistics processes (i.e., cold storage, packaging, loading, unloading, transport, and distribution) determine the quality and performance of the cold chain. It is necessary for cold chain professionals to have a general understanding on how these facilities work to maintain an intact and cost-effective cold chain.

4.3.1   Refrigerated Vehicles

For refrigerated road transport, refrigerated vehicles are widely used because of its flexibility and support for door-to-door transport. Unlike factory or warehouse where the temperature is easily controlled and monitored, refrigerated vehicles can be a crucial point of the whole chain, since products may undergo transient conditions during the loading operations or even during the whole transport process, with great risks for the good integrity (Carullo et al. 2009). Therefore, to follow specified temperature guidelines for goods in storage during transport, installing an effective management system for monitoring and controlling is essential.

Many factors should be considered in the design of a refrigerated transportation unit: extremes of exterior weather conditions, desired interior conditions, insulation properties, infiltration of air and moisture, trade-off between construction cost and operating costs, and physical deterioration from shocks and vibrations (Tassou et al. 2009). Cost, productivity, and efficiency are main factors that need to be considered in the design of the refrigerated vehicle.

A refrigerated vehicle is always heavier than a conventional trailer, and it consumes more fuel to pull the extra weight. To enable the vehicle to operate with fuel economy and have more capacity, aluminum or other lightweight material is used instead of steel. Dual tires are replaced by wide-base single tires in order to save fuel, and thermal insulation is designed according to the application requirement (Gelinas 2007). The environment inside the cargo compartment may be temperature controlled or temperature modified.

Temperature-controlled vehicles include vans, rigid trucks, and semi-trailers that have an insulated thermostatically controlled cargo compartment and a dedicated refrigeration unit capable of maintaining the labeled temperature range of the products being transported. Vans and small rigid trucks typically have refrigeration units powered directly by the vehicle's engine. Larger rigid vehicles and semi-trailers have independent diesel-powered refrigeration units. Both types may also have electrical backup so that they can be powered while parked.

Figure 4.3 shows a temperature-controlled truck which utilizes the power of the sun to charge Transport Refrigeration Unit (TRU) batteries to maintain peak performance in an environmentally sustainable way using a solar photovoltaic (PV) refrigerated vehicle.

Temperature-modified vehicles are similar to temperature-controlled refrigerated vehicles, except that the vehicle itself simply moderates the ambient temperature, either by heating or by cooling. The transported product is generally packed in a qualified passive shipping system designed to keep it within the labeled temperature range. The temperature-modified environment in the vehicle serves to extend the autonomy of the passive shipping system and protect the product from temperature extremes. Care must be taken not to subject the packages to refrigerated temperatures (e.g., +2 to +8 °C for pharmaceuticals) for extended periods because this risks freezing the package contents. Therefore, all refrigerated vehicles should be equipped with an on-board electronic temperature-monitoring and event logger system (WHO 2011).

Fig. 4.3 Temperature-controlled vehicle with solar power (Reprinted by permission of http://prod. sandia.gov)

Cold plate is the central part that maintains the product in a specified temperature range. Traditionally, the cold plate in the refrigerated vehicle is powered by fuel. An advanced cold-plate system is powered by electricity and can be recharged at night. A six to eight hours charge may power the cold plate for more than 12 h with up to 48 h of product protection time. This innovative electricity-powered cold-plate system may save 80% fuel cost comparing with the traditional fuel-powered one. Additionally, it produces no diesel pollution (Gelinas 2007).

Demands for mixed loads of products require different storage temperatures, and the trend of refrigerated transport is to use multi-compartmental vehicles (Kuo et al. 2005; James et al. 2006). Compartments are divided by movable or fixed insulated panels longitudinally. Half-width horizontal insulated bulkheads that consist of compression-fit foam with a vinyl outer layer are used to divide trailers into multi-temp zones. While the reefer is not fully loaded, a bulkhead can be adjusted to the utilized capacity to save fuel (Klie 2005). The advantage of using the multi-compartmental vehicle is flexibility and space-saving, but operating procedures are more complicated, and this also affects costs. The cost of a multi-compartment temperature-controlled vehicle is about £100,000 compared to around £30,000 for an ambient one (Smiths and Sparks 2004). Figure 4.4 shows the concept of a multi-temperature trailer.

In Europe, the specifications for refrigerated vehicles are covered by the Agreement on the International Carriage of Perishable Foodstuffs and on the special equipment to be used for such carriage (ATP). There are standards for refrigerated equipment for the carriage of perishable foodstuffs (Estrada-Flores and Eddy 2006). The agreement states that new refrigeration equipment installed on a refrigerated vehicle must have a heat extraction capability at the class limit temperature of at least 1.35 times the heat transfer through the walls in a 30 °C ambient temperature and 1.75 times if the refrigeration unit was tested separately outside the vehicle to determine its

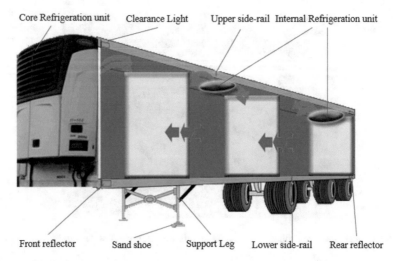

Core Refrigeration unit   Clearance Light   Upper side-rail Internal Refrigeration unit

Front reflector   Sand shoe   Support Leg   Lower side-rail   Rear reflector

Fig. 4.4 Multi-temperature trailer

effective cooling capacity at the prescribed temperature. The ATP certificate ensures that the insulated body and the refrigeration unit have been tested by a third party and that the two have been appropriately matched (Tassou et al. 2009).

Figure 4.5 shows a concept of smart or intelligent container. As shown in the figure, future container will be equipped with various Internet of Things (IoT) devices such as sensors, Radio Frequency Identification (RFID), and communication devices.

A cold-trace trailer was developed during cold-trace project (Ursa et al. 2006). Cold trace provides fleet managers with real-time information necessary to ensure the safe, prompt, and efficient delivery of goods in the most cost-effective way possible. It is a management tool on board the trucks that collects and processes information from a set of sensors distributed throughout the vehicle. The tool records information every 20 s, sends information every minute, and manages all administrative tasks performed during the trip.

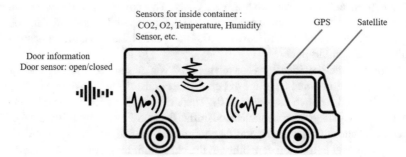

Sensors for inside container :
   CO_2, O_2, Temperature, Humidity
Sensor, etc.

GPS   Satellite

Door information
Door sensor: open/closed

Fig. 4.5 Intelligent container

4.3.2 Cold Store/Refrigerated Warehouse

Cold stores (Europe) or refrigerated warehouse (US) is facilities for handling and storing perishables under controlled temperatures in order to maintain product quality (Duiven and Binard 2002).

The requirement for cold storage/refrigerated warehouse is growing as the demand for fresh, refrigerated, and frozen food is increased. According to report from International Association of Refrigerated Warehouses (IARW), the total capacity of refrigerated warehouses is estimated at 552 million cubic meters worldwide in 2014, an increase of 92 million cubic meters (20%) over 2012 (IARW 2014).

A refrigerated warehouse is capital-intensive with high building and equipment costs, generally more than two times the cost of a conventional one (Sethi 1999). Therefore, the challenge of designing cold storage facilities is to ensure accurate control of the environment under the lowest energy consumption (Merli 1999). Facilities inside the refrigerated warehouse (e.g., racking system, rollers/conveyors, refrigerated dock design, etc.) determine the operation efficiency and space usage factor in it.

Refrigerated warehouse facilities operate in different fashions, depending upon whether they offer public or private refrigerated space. Public general storage facilities typically store food for clients at a stated unit rate. Private general storage facilities exist to facilitate an operator's role—often that of a producer, processor, or manufacturer of refrigerated food products. Semi-private facilities store an operator's products in addition to offering storage space to outside clients. All facilities attempt to turn product over quickly, aiming for "just-in-time" delivery. Most refrigerated warehouse facilities have loading docks, and nearly all have interiors divided into cooler space and freezer space. Cooler space temperatures may range from 0 to 50 °F, while freezer space temperatures range from -5 to -30 °F. Among all operators, freezer space occupies 78% of total warehouse area; cooler space fills the remaining 22% (Gottlieb 2006). Interior view of a refrigerated warehouse is shown in Fig. 4.6.

4.3.3 Work in Refrigerated Warehouse in Cold Chain

Occupational Safety and Health Administration (OSHA) under US Department of Labor published various guidelines for cold chain work environments. According to 1970 Occupational Safety and Health Act (OSHA 1970), it shall be the duty of an employer to ensure the health, safety, and welfare at work of all their employees. Some are as follows:

- Employers should train workers. Training should include:

  - How to recognize the environmental and workplace conditions that can lead to cold stress.

Fig. 4.6 Interior view of a cold storage warehouse

- The symptoms of cold stress, how to prevent cold stress, and what to do to help those affected.
- How to select proper clothing for cold, wet, and windy conditions.

Employers should:

- Monitor workers' physical condition.
- Schedule frequent short breaks in warm dry areas, to allow the body to warm up.
- Schedule work during the warmest part of the day.
- Use the buddy system (work in pairs).
- Provide warm, sweet beverages. Avoid drinks with alcohol.
- Provide engineering controls such as radiant heaters.

Working in low temperatures can cause accidents easily because the body of workers is prone to low temperature, and for this reason, biological reaction, and intelligence and work performance are degraded.

There are a few research on the measurement of worker's biological reaction and work performance in low temperature. Repetitive works in low temperature cause the biological changes, and this could be measured by skin temperature, rectal temperature, heart bit, pressure, trembling in hands. To measure the decrease in work performance, ring test, counting task, attention test, pinch strength, and hand-grip strength were measured. Some research proved that there are correlations between biological change and work performance (Tochihara et al. 1995; Kim et al. 2007).

"Cold Stress Guide" by OSHA recommended a work-warming schedule considering a work shift change of every 4 h as in (OSHA 2019). The guideline suggests maximum work period (i.e., time) and number of break (each break is 10 min worm-up period) considering wind speed and air temperature, and mechanical and manual processes must be designed accordingly.

Despite a lot of effort for protecting workers' health and improving work performances, most cold work environments are not good for workers. With the fast development of automation technologies and the fast adoption of Industry 4.0 in warehouse, automation in cold chain is inevitable trend.

4.3.4 Automation in Cold Chain Environment

Each node along the cold supply chain has different operational environment. Some requires room temperature, while other environment requires low temperature (e.g., $-10\,°C$, etc.). Most cold chain operation environment is not good for man and facility. For example, forklift operating below $0°$ has impact on the battery life.

To address such issue, warehouses are investing in a few types of automated storage—power roller-based automated transferring system, automated storage and retrieval systems systems, and mobile racking—to keep workers warm, reduce energy, and increase productivity (Lewis 2018). Power roller-based automated system is a system for moving a product from A to B. Since cold chain process requires waterproof device and frequent cleaning with cold or cool temperature, it should be robust against such condition. Figure 4.7 shows examples of the power rollers system used in cold chain environment.

Whether semi- or fully automated, cold storage facilities can automate to cut costs, improve productivity, and achieve compliance. However, a few things should be considered for facility automation in cold chains:

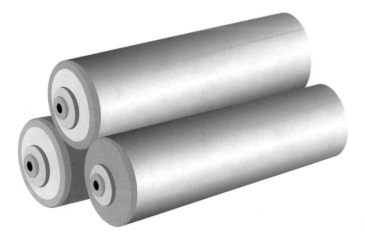

Fig. 4.7 Power roller system samples

- Operation: Automated facilities should have the same lifecycle in cold environment. Electrical wiring, lubrication must be freezer rated.
- Maintenance: Because a maintenance operation is performed in normal temperature, electronic devices could corrode due to condensation (the temperature change from cold temperature to normal temperature) and lifecycle could be reduced.
- Cleaning: Cleaning could be considered as a maintenance action. In case of facilities dealing with food, high pressure and temperature cleaning is typical. Facilities and components used in such environment should have strong endurance and reliability.

4.3.5   Refrigerated Containers/Reefers

One of the best solutions to transporting food products in the regions with high-temperature exposures is the use of refrigerated containers. These containers feature extremely heavy insulation.

There are two basic types of refrigerated container (reefer): the porthole and integral type (Fig. 4.8). Porthole refrigerated containers, also known as insulated or conair containers, do not have their own refrigeration unit and rely on an external supply of cold air (such as central refrigeration plant). The lack of a refrigeration unit allows to have a larger internal volume and payload than integral units. Porthole containers are used with two sealable portholes on the front bulkhead, through which refrigerated air is circulated. Integral refrigerated containers, however, have their own integrated refrigeration unit. This is generally electrically powered and involves a three-phase electric power supply (GDV 2005).

Integral containers are equipped with a refrigeration plant and need merely to be supplied with electricity (three-phase, 380 V/50 Hz, or 440 V/60 Hz) to maintain their internal environment. This configuration is more widely used. Integral containers are generally carried on deck of ships in marine transport: However, modern ships include

Fig. 4.8 Refrigerated container (integral)

space below deck for these containers with additional air renewal to remove heat from the container's condensers. Some ships are also fitted with water-cooling facilities for containers with water-cooled condensers, although not all heat is removed, and ventilation is still necessary. Those carried on deck may be protected from the sun with an upper layer of non-refrigerated containers (IIF/IIR 1995).

In the transport phase, temperature-controlled products are preserved in refrigerated containers. The development of mechanical refrigeration, controlled atmospheres (CA), and packaging provides solid technical support for international trade of various products under temperature-controlled conditions (James et al. 2006).

The refrigerated containers are insulated and equipped with refrigeration units in their structures. The units are powered by electricity from an external power supply either on board the ship or from a generator on vehicle. The container is connected to the ship's refrigerated system, and temperature is easy to control. It is crucial to make sure the refrigeration unit is running all through the journey (James et al. 2006).

Apart from temperature control, controlled atmosphere (CA) technology is designed to preserve the freshness of the postharvest during shipping. The system maintains a balanced atmosphere of oxygen and other gases such as N_2, to minimize respiration in postharvest. CA technology can also improve the control of insects in some commodities and prevent water loss and weight shrinkage (Sowinski 1999).

It is noteworthy that a refrigerated container cannot cool down a product. The function of it is to maintain the current temperature of a product to ensure the product integrity (McGovern 1998). In other words, it is important to ensure both the product and refrigerated container are at the right temperature before loading (James et al. 2006).

4.3.6 Cold Chain Equipment

Procuring the needed equipment is one aspect of keeping a functional cold chain. To be effective, cold chain equipment must be properly managed and maintained. The usual types of cold chain equipment are refrigerator (or freezer) and cold boxes.

(1) **Refrigerator or Freezer**

Refrigerator could be typically classified into three types:

– **Bar Fridge Units**

Any style of small, single door (bar style) fridge is not recommended for the storage of vaccines and biologics. This type of fridge is unpredictable and may not maintain temperatures necessary for product stability.

– **Domestic (Home) Refrigerators**

Domestic combination refrigerator and freezer units, though not recommended for storage of vaccines and biologics, are acceptable. Domestic refrigerators are designed for food storage and not for storage of vaccines and biologics. Precautions and fridge modifications are needed. IoT technology and high level of user requirement have driven the global smart refrigerator market recently. Smart refrigerator can identify what kinds of products are being stored and keep a track of the stock through automatic identification technologies. Even though the smart refrigerator market is expected to grow in the near future, but because of its high price, it is not adopted very fast. In Chap. 9, we presented a design and development of a smart refrigerator.

– **Purpose-Built (Commercial) Refrigerators**

A purpose-built (commercial) vaccine refrigerator (also referred to as pharmacy, laboratory style, or laboratory-grade refrigerator) is the standard for storing large inventories of vaccines and biologics. The advantage is that they are specifically designed to ensure better temperature regulation (CDC 2013).

In using the domestic freezers, the "star rating" of the freezer is considered; a three-star freezer is capable of temperatures below –18 °C, a two-star freezer of temperatures below –12 °C, and a one-star freezer of temperatures below –6 °C (Martin and Ronan 2000).

Blood refrigerators, plasma freezers, and platelet agitators are the blood cold chain equipment used for the storage of blood components. The purpose of a blood refrigerator is to store whole blood and red cells at between +2 °C and +6 °C. There are various compression-type blood refrigerators using CFC-free refrigerant gas for use in different environments. They are generally frost and condensation free.

"Compression-type" plasma freezers are suitable for the storage of Fresh Frozen Plasma (FFP) and cryoprecipitate. The main difference between a blood refrigerator and a plasma freezer is in the temperatures that they are capable of maintaining. A plasma freezer is expected to operate at a temperature of below −30 °C. The equipment should use CFC-free refrigerant gas and electricity supply from the national grid. The freezer has an internal fan cooling mechanism to ensure the even distribution of air in the cabinet.

Platelet concentrates are harvested from whole blood by centrifugation or during platelet apheresis. Platelet concentrates are suspended in about 60 ml of plasma. The packs are continually agitated in a platelet agitator in a room with an ambient temperature of between +20 and +24 °C. This generally requires that the laboratory is air-conditioned in order that the temperatures are maintained within the desired range. The recommended type of agitator is a flatbed agitator with horizontal or vertical agitation as this ensures no platelet clumps are formed (WHO 2005).

(2) **Cold Boxes**

Cold boxes are insulated containers that can be lined with coolant packs to keep fresh and frozen products cold during transportation and/or short-period storage. They are also used to temporarily store fresh products when the refrigerator is out of order or being defrosted. Figure 4.9a & b show examples of different cold boxes developed lately.

In case of vaccine storage, the capacity of cold boxes is between 5.0 and 25.0 L. There are two types of cold boxes:

a

b

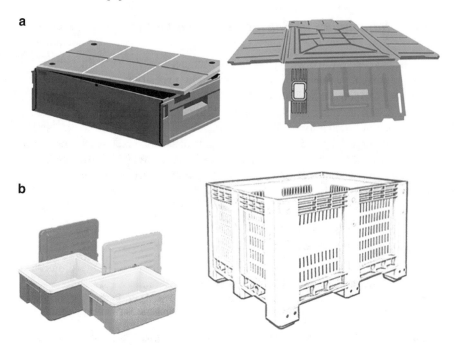

Fig. 4.9 a RFID embedded box. **b** Cold box and container for agriculture products

- short range: with a minimum cold life of 48 h;
- long range: with a minimum cold life of 96 h (UNICEF 2014).

In contrast, blood transport boxes must be specially designed to maintain the internal temperature between +2 and +10 °C for at least 24 h using appropriate ice packing, i.e., to have a cold life of at least 24 h. Transport boxes are manufactured in different sizes to suit different needs. The manufacturer of the transport box defines the number of ice packs required to keep blood within a temperature range of +2 to +10 °C. However, the quantity and type of coolants to be used will depend on the blood components or products to be transported and the distance.

It is important to highlight a major difference between the use of blood transport boxes and that of vaccine cold chain transport boxes that affects the design of the boxes. Although vaccine transport boxes are often used to transport blood, vaccine cold boxes are designed to be able to transport products over a period of up to five days, especially to reach remote villages. Blood transportation is limited to fairly short periods, generally below 24 h (WHO 2005).

4.3.7   Maintenance of Refrigeration Equipment

The intended design life of cold stores can be achieved only if a considered program of regular preventive maintenance has been put in place. There are two elements of a cold store that require maintenance: the refrigeration equipment and the insulated envelope. Close attention should be paid to both of these elements. In order to minimize the risk of product loss, most cold stores should have a duplicate refrigeration system, an emergency power supply, and a sophisticated temperature-monitoring and alarm system, all of which also need to be maintained (WHO 2014). The maintenance schedule can be daily or weekly or monthly or yearly based on the types of equipment and its functions.

References

Ames H (2006) Authentication from a cold chain perspective. Pharmaceutical Commerce

Barbosa-Cánovas GV, Altunakar B, Mejía-Lorío DJ (2005) Freezing of fruits and vegetables—an agribusiness alternative for rural and semi-rural areas. FAO Agric Serv Bull 158, ISBN 92-5-105295-6

Brecht JK, Sargent SA, Brecht PE, Saenz J, Rodowick L (2019) Protecting perishable foods during transport by truck and rail, USDA handbook No.669 (revised), IFAS extension, University of Florida

Carullo A, Corbellini S, Parvis M, Reyneri L, Vallan A (2009) A measuring system for the assurance of the cold-chain integrity. In: Proceedings of the IEEE international instrumentation and measurement technology conference, Vancouver, Canada, pp 1598–1602

Casper C (2007) Safety starts with temperature control: latest technologies enhance product quality and safety (cold chain report). Food Logist 12:16–20

CDC (2013) Cold chain protocol—vaccines and biologics, communicable disease control, Manitoba, Canada

Crisosto CH, Mitchell FG, Johnson RS (1995) Factors in fresh market stone fruit quality. Postharvest News Inform 6(2):17N–21N

Coulomb D (2008) Refrigeration and cold chain serving the global food industry and creating a better future: two key IIR challenges for improved health and environment. Trends Food Sci Technol 19:413–417

Delgado AE, Sun DW (2000) Heat and mass transfer for predicting freezing process, a review. J Food Eng 47:157–174

Duiven JE, Binard P (2002) Refrigerated storage: new developments. International Institute of Refrigeration

Estrada-Flores S, Eddy A (2006) Thermal performance indicators for refrigerated road vehicles. Int J Refrig 29:889–898

Fennema OR, Powrie W, Marth EH (1973) Low temperature preservation of foods and living matter. CRC Press

GDV (2005) Container handbook—cargo loss prevention from German marine insurers. German Insurance Association. Gesamtverband der Deutschen Versicherungswirtschaft E.V. www.contai nerhandbuch.de

Gelinas T (2007) Cooling cost control. Fleet Equipm 33(2):22–27

George RM (1993) Freezing process used in food industry. Trends Food Sci Technol 4:134

Gottlieb MS (2006) Refrigerated warehousing: an industry study, MSG accountants, consultants, and business valuators

IARW (2014) 2014 IARW global cold storage capacity report

IATA (2011) Perishable cargo regulations, 11 the edn. International Air Transport Association, Monteral-Geneva

IIF/IIR (1995) Guide to refrigerated transport. International Institute of Refrigeration, Paris

IIR (2007) Refrigeration drives sustainable development: state of the art- report card. International Institute of Refrigeration

IIR (2008) Refrigeration and cold chain serving the global food industry_IIR challenges. Trends Food Sci Technol 19:413–417

IIR (2009) The role of refrigeration in world nutrition. 5th informatory note on refrigeration and food. International Institute of Refrigeration

IIR (2015) The role of refrigeration in the global economy. 29th informatory note on refrigeration technologies. International Institute of Refrigeration

IIR (2019) The role of refrigeration in the global economy. 38th informatory note on refrigeration technologies. International Institute of Refrigeration

IIR (2020) The role of refrigeration in world nutrition. 6th informatory note on refrigeration and food. International Institute of Refrigeration

James SJ, James C (2014) Chilling and freezing of foods. Book chapter in food processing: principles and applications, 2nd edn. Wiley, pp 79–105

James SJ, James C, Evans JA (2006) Modelling of food transportation systems—a review. Int J Refrig 29:947–957

Johnston WA, Nicholson FJ, Roger A, Stroud GD (1994) Freezing and refrigerated storage in fisheries. Fisheries technical paper-340, FAO

Kader A (2002) Postharvest technology of horticultural crops, 3rd edn. University of California. Agricultural and Natural Resources. Publication 3311

Kim TG, Tochihara Y, Fujita M, Hashiguchi N (2007) Physiological responses and performance of loading work in a severely cold environment. Int J Ind Ergon 37(9–10):725–732

Kitinoja L (2013) Use of cold chains for reducing food losses in developing countries. Postharvest Education Foundation (PEF) White paper No. 13-03

Klie L (2005) Cool only what's necessary. Frozen Food Age 54(2):33–34

Kuo J-C, Chen M-C (2010) The development of multi-temperature joint distribution system for the food cold chain. Food Control 21(4):559–566

Kuo J-C, Chen M-C, Chang F-Y (2005) The development of multi-temperature joint distribution system for effective logistics model. In: International Association for Management of Technology (IAMOT), Vienna, Austria

Lewis C (2018) Cold chain embraces automated storage. Modern materials handling. https://www.mmh.com/article/cold_chain_embraces_automated_storage. Accessed 21 Apr 2019

Martin G, Ronan G (2000) Managing the cold chain for quality and safety. FlairFlow Europe technical manual 378A/00

Martinez-Romero D, Bailen G, Serrano M, Guillen F, Valverde JM, Zapata P, Castillo S, Valero D (2007) Tools to maintain postharvest fruit and vegetable quality through the inhibition of ethylene action: a review. Crit Rev Food Sci Nutr 47(6):543

McGovern JM (1998) Cool technology. Transp Distrib 39(12):25–28

Mercier S, Villeneuve S, Mondor M, Uysal I (2017) Time-temperature management along the food cold chain: a review of recent developments. J Food Sci 16:647–667

Merli R (1999) PRWs seeking new ways to improve efficiency. Frozen Food Age 48(1):46–47

OSHA (1970) OSH Act of 1970. https://www.osha.gov/laws-regs/oshact/toc. Accessed 21 Apr 2019

OSHA (2019) Cold stress guide. https://www.osha.gov/SLTC/emergencypreparedness/guides/cold.html. Accessed 21 Apr 2019

R2N (2003) Expired product project, developed for the Joint Industry Unsaleables Steering Committee, prepared by Reftery Resource Network, Inc.

Sahin E, Babaï MZ, Dallery Y, Vaillant R (2007) Ensuring supply chain safety through time temperature integrators. Int J Logist Manag 18(1):102–124

Salin V, Nayga RM (2003) A cold chain network for food exports to developing countries. Int J Phys Distrib Logist Manag 33(10):918–933

Saltveit ME (2016) Respiratory metabolism. The commercial storage of fruits, vegetables, and florist and nursery stocks. In: Gross KC, Wang CY, Saltveit M (eds) Handbook number 66, revised version. United States Department of Agriculture, Agricultural Research Center

Sethi S (1999) Building win-win. Frozen Food Age, pp 24–27

Smith D, Sparks L (2004) Temperature controlled supply chain. In: Bourlakis MA, Weighman PWH (eds) Food supply chain management. Blackwell Publishing, pp 179–198

Sowinski L (1999) Keep your big deal from melting away: shipping perishables call for efficiency and expertise. World Trade 12(3):70–72

Stoecker WF (1998) Refrigeration and freezing of food. In: Industrial refrigeration handbook. McGraw Hill

Tassou SA, De-Lille G, Ge YT (2009) Food transport refrigeration—approach to reduce energy consumption and environmental impacts of road transport. Appl Therm Eng 29:1467–1477

Thompson JF, Kader AA (2001) Wholesale distribution centre storage, perishable handling quarterly, p 107. Retrieved from http://ucce.ucdavis.edu/files/datastore/234-102.pdf. Accessed 26 Aug 2021

Tochihara Y, Ohnaka T, Tuzuki K, Nagai Y (1995) Effects of repeated exposures to severely cold environments on thermal responses of humans. Ergonomics 38(5):987–995

UNICEF (2014) Procurement guidelines, vaccine carriers and cold boxes. http://www.unicef.org/supply/files/Vaccine_Carriers_and_Cold_Boxes.pdf. Accessed 25 Apr 2020

Ursa Y, Perez P, Meissner A (2006) Cold-trace: a mobile-based traceability solution rendering fleet management more effective. In: Cunningham P, Cunningham M (eds) Exploiting the knowledge economy: issues, applications, case studies. IOS Press, Barcelona, Spain

Wang CY, Wallace HA (2003) Chilling and freezing injury. In: Agriculture Handbook Number 66, U.S. Department of Agriculture

WHO (2005) Manual on the management, maintenance and use of blood cold chain equipment. World Health Organisation

WHO (2011) Temperature-controlled transport operations by road and by air. Technical supplement to WHO technical report series, No. 961

WHO (2014) Maintenance of refrigeration equipment. Technical supplement to WHO technical report series, No. 961

Yuzainee MY, Noor Zafira NH, Nurulnadiah J (2019) Cooling load calculation for efficient cold storage of fresh-cut yam bean. Int J Recent Technol Eng (IJRTE) 8(4):6506–6513. ISSN: 2277-3878

Zhang L (2007) Cold chain management. Master's thesis. Cranfield University, UK

Chapter 5
Cold Chain Monitoring Tools

Since a cold chain refers to a temperature-controlled supply chain, data collection (e.g., product tracking, temperature tracking) methods are vital to maintain a sustainable and unbroken cold chain. Selecting the right temperature-monitoring tool is critical to avoid spoilage, to ensure pinpoint accuracy and to comply with mandatory regulations. In general, there are five options of tools to monitor cold chain equipment/facilities such as fridge, freezer, cold storage, and warehouse:

- thermometers/probe thermometers;
- chart recorders;
- temperature indicator labels;
- data loggers;
- wireless sensors/Internet of Things (IoT) devices [radio frequency identification (RFID), wireless sensor network (WSN), etc.].

5.1 Thermometers

A thermometer is an instrument that measures the internal temperature of cold chain equipment such as refrigerators, cold boxes, and vaccine carriers. Thermometers have been used to check fridge and freezer temperatures for ages and are the most basic instrument available. A technician or other team member usually checks each fridge or freezer at least twice a day and records the reading from a thermometer that monitors the internal temperature. The first widely used thermometer is mercury-in-glass thermometers, but there are other types available today.

A probe thermometer is one of the most used temperature-measuring tools which are small, easy to use, and provide instant readings for various products in cold chain. Probe thermometers have an advantage over surface thermometers in that they can provide an exact core temperature by reaching the center of the food item. Figure 5.1 shows the examples of mercury-in-glass thermometer and probe thermometer.

© Springer Nature Switzerland AG 2023
M. M. Aung and Y. S. Chang, *Cold Chain Management*, Springer Series in Advanced Manufacturing, https://doi.org/10.1007/978-3-031-09567-2_5

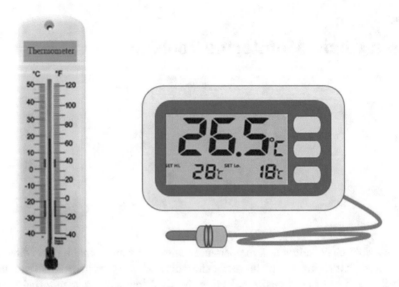

Fig. 5.1 Mercury-in-glass thermometer and probe thermometer

The above tools that can be destructive or non-destructive type typically use manual recordings that are less reliable in the event of a regulatory or liability action. These simple instruments have the benefits of low capital costs, ease of use, and a long history, but they also have certain disadvantages, such as high labor costs, burden of record keeping, risk of human error, and inability to do continuous monitoring.

5.2 Chart Recorder

Chart recorder is an electromechanical piece of equipment that documents/records a mechanical or electrical input signal or trend onto a chart, paper, or a rolled piece of paper (Fig. 5.2). Chart recorders are built in three primary formats (https://en.wikipe dia.org/wiki/Chart_recorder). Strip chart recorders have a long strip of paper that is ejected from the recorder. Circular chart recorders have a rotating disk of paper that must be replaced more often, but are more compact and amenable to being enclosed behind glass. Roll chart recorders are similar to strip chart recorders except that the recorded data are stored on a round roll, and the unit is usually fully enclosed. There are different types of chart recorders to record environmental conditions such as temperature and humidity during storage and transportation of food, pharmaceutical, chemicals, and other environmentally sensitive commodities.

Chart recorders are simple to use with an uncomplicated form of data storage and can offer continuous monitoring. But, the operating cost is relatively high over time, and the method of record keeping is cumbersome. Figure 5.2 shows a temperature and

Fig. 5.2 Chart recorder device

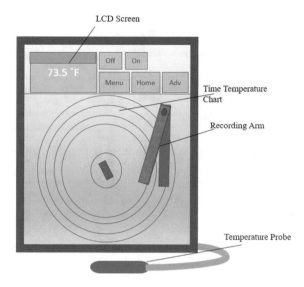

humidity circular chart recorder. Recently, paperless chart recorders have become more widely used than paper-based chart recorders.

5.3 Time–Temperature Indicator

Temperature indicator (TI) and time–temperature indicator (TTI) typically refer to temperature-sensitive color-changing labels. TI can provide a permanent record of temperature abuse (i.e., they provide a non-reversible record of temperature exposure) and indicate simply by colors if the temperature is above or below the specified interval. Temperature indicator provides a highly visible, easy to read indication of when your product has been exposed to temperatures that are too hot or too cold during transport or storage. Figure 5.3 shows an example of temperature indicator.

Fig. 5.3 Example of temperature indicator

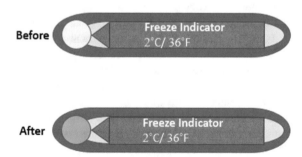

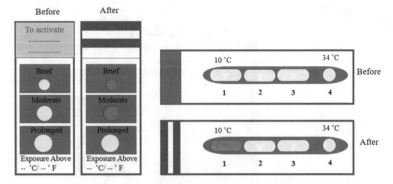

Fig. 5.4 Examples of time–temperature indicator

TTI is based on chemical, physical, or microbiological reactions and can indicate quality problems with a color code based on the accumulated time and temperature history of a product (Fig. 5.4). Time–temperature indicators monitor product temperature exposure through the entire supply chain during transportation and storage (e.g., frozen or refrigerated foods, drugs, vaccines, medical diagnostic kits, blood, blood products, and intraocular lenses). They provide a non-reversible record of temperature exposure that is accurate and easy to interpret.

In the cold chain for perishable products, an unpredictable temperature condition could be occurred frequently, and such condition could result in high product losses. Customer needs evidence of good product quality across their supply chains for quality assurance and to prepare unexpected insurance claims.

5.4 Data Loggers

A data logger device calculates the product's quality based on time and temperature and visualizes the result with an light-emitting diode. In contrast to the TTI, it can be used multiple times and allows the temperature history to be read out through a serial interface. They provide continuous freezer/fridge temperature monitoring and can trigger an alert if temperatures go outside of acceptable ranges (Fig. 5.5).

Many foods, dairy, and pharmaceutical companies are already monitoring and tracking their environmentally sensitive products using temperature data loggers placed in their transportation vehicle, containers, or even pallets. But, these are usually expensive and not automated, thus require manual inspection. In order to read the temperature information recorded, it is necessary to open the container or package containing the food and read temperature information at the final destination. For these reasons, their use is limited only to some parts of the cold chain or to a few types of products, while for other products and important parts of the chain, continuous product temperature monitoring is not completed (Syntax Commerce

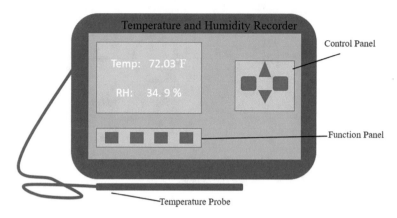

Fig. 5.5 Data logger device

2007; Abad et al. 2009). Recently, data logger can be installed in the container and stay in the storage.

5.5 Radio Frequency Identification and Sensors

Radio Frequency Identification (RFID) is an automatic identification method, relying on storing and remotely retrieving data using devices called RFID tags or transponders. RFID system often implies the process and physical infrastructure by which a unique identifier within a predefined protocol definition is transferred from a device to a reader via radio frequency waves. It has taken many years of development to come up with a functional system, but the basic principle is not much different than that of the well-known bar code: Encode an identifier number in a machine-readable form that can be accessed quickly and reliably with no human translation. However, it is not fair to say that RFID is just a glorified bar code transferred via radio frequency waves. The very nature of RFID, the fact that it is based on a microprocessor containing a data memory space, allows RFID chips to be applied in many instances that could not have been ever imagined with bar codes. Figure 5.6 shows extended architecture for RFID (Banks et al. 2007). As in the figure, tags and the antenna of reader (i.e., interrogator) communicate based on the specific protocol.

An RFID tag (or transponder) is an object that can be applied to or incorporated into a product, animal, or person for the purpose of identification using radio waves. The tags can be read by readers from several meters away without line of sight. Most RFID tags contain at least two parts. One is an integrated circuit for storing and processing information, modulating and demodulating a radio frequency (RF) signal, and other specialized functions. The second is an antenna for receiving and transmitting the signal. Semi-passive RFID tags or active RFID tags equipped with

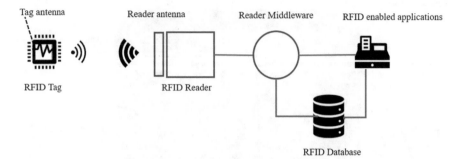

Fig. 5.6 Radio frequency identification infrastructure

a temperature sensor allow the temperature history to be read out through a RF interface (Illic et al. 2009). Figure 5.7 shows sample RFID tags.

An RFID reader (or interrogator) is a network connected device (fixed or mobile) with an antenna that sends power as well as data and commands to the tags. An RFID reader is basically a RF transmitter and receiver, controlled by a microprocessor. Figure 5.8 shows an example of RFID reader.

Frequency refers to the size of the radio waves used to communicate between RFID reader and tag. Radio waves behave differently at each frequency. If an RFID system operates at a lower frequency, it has a shorter read range and slower data read rate, but increased capabilities for reading near or on metal or liquid surfaces. If a

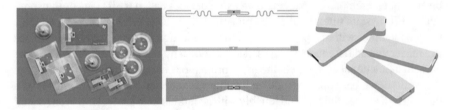

Fig. 5.7 Radio frequency identification tags

Fig. 5.8 Radio frequency identification reader: fixed reader with antennas

Table 5.1 Radio frequency identification system applications by frequency

Frequency	Distance	Applications
125–135 kHz	10 cm or less	Animal tracking, rental item, work in process (WIP), auto-immobilizers, etc.
13.56 MHz	10 cm–1 m	Security/access control, bus card, auto-immobilizers, rental item, WIP, etc.
865–868 MHz (Europe) 902–928 MHz (North America) UHF	3–6 m	Supply chain management, asset management, baggage tracking, toll collection, yard management, etc.
UHF (433 MHz)	100 m	Real-time location system, asset management with temperature monitoring, yard management, E-seal, etc.
2.45 GHz	a few meters	Real-time location system, asset management, supply chain management, location tracking, E-seal, etc.

system operates at a higher frequency, it generally has faster data transfer rates and longer read ranges than lower frequency systems, but more sensitivity to radio wave interference caused by liquids and metals in the environment (http://www.impinj.com). Table 5.1 shows RFID frequencies with their potential applications. A supply chain is a multi-level concept that covers all aspects from raw material to a final product. In terms of life cycle, it is from shipping to a point of sale, followed then by use/maintenance, and then followed potentially by disposal/or returned goods. The ISO/IEC 17363 through ISO/IEC 17367 series was prepared by the Joint Working Group (JWG) of ISO Technical Committee TC 122 and ISO Technical Committee TC 104. Figure 5.9 shows RFID standards for supply chain management. These standards pertain to supply chain applications of RFID (ISO/IEC 17364 2013). The series includes the following:

- ISO/IEC 17363, supply chain applications of RFID—freight containers;
- ISO/IEC 17364, supply chain applications of RFID—returnable transport items;
- ISO/IEC 17365, supply chain applications of RFID—transport units;
- ISO/IEC 17366, supply chain applications of RFID—product packaging;
- ISO/IEC 17367, supply chain applications of RFID—product tagging.

RFID uses electromagnetic fields to automatically identify and track tags attached to objects. RFID technologies can improve the performance of cold chain through the following uses:

- as a mean to identify items through a unique electronic product code;
- as a mean to track the geographical position of individual packages, pallets, shipping containers, or trucks, which can be stationary or in movement during distribution.

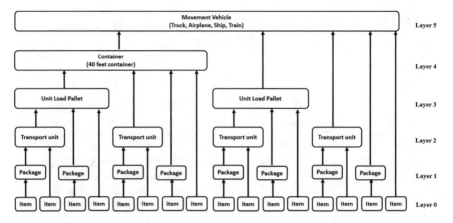

Fig. 5.9 Radio frequency identification standard with supply chain (Reprinted by permission of Alien Technology LLC)

While RFID is typically used in track and trace applications, sensors are ideally suited to monitor environmental parameters that could be degraded over time (e.g., as a result of temperature, humidity, etc.). In the case of perishable goods, there is generally a need to monitor goods continuously, since there could be various changes which impact on the quality: a degradation of quality, creation of a hazard, reduction of value, or reduced shelf life, etc. These degradations are associated with continuous variables and are often for more fundamental parameters such as temperature or humidity. However, other parameters can also be sensed to indicate the presence of gases, etc.

The deterioration of perishable objects can lead to a decrease in the esthetic appeal (such as discolorations, blemishes, and unusual smells), as well as a reduction in nutritional value (e.g., due to loss of vitamins) and potentially the production of poisonous toxins, such as those produced by the bacteria *Salmonella* and *Escherichia coli*. Sometimes, the degradation of foods is readily visible as changes of texture or discoloration, such as the blackening of banana skins as the fruit ripens, then passes peak ripeness. In other situations, the degradation might not be so readily visible, even though the efficacy of the object has been degraded.

While the human senses have only a limited capability to assess the intrinsic product properties, modern sensor technologies can help to provide the required information. Tracking environmental parameters such as temperature and vibration for individual logistic units allows problems in the supply chain to be spotted and the actual quality levels of individual products to be more precisely predicted (Sahin et al. 2007), all of which aids the decision-making with respect to the future distribution of products.

Sensor-based tracking systems or WSNs are really powerful not only to measure direct environmental conditions (i.e., temperature, humidity) before (and during) deterioration but also to indicate the extent of progress or rate of degradation (i.e.,

Table 5.2 Key environmental factors and related sensor types (Reprinted by permission of Bridge Project, Bowman et al. 2009)

Factor	Sensor type for detection of preconditions
Gases such as oxygen, carbon dioxide, ethylene	Gas sensors
Relative humidity	Humidity sensors
High temperatures	Temperature sensors (maximum exposure)
Temperature fluctuations	Temperature sensors (maximum, minimum, history)
Low temperatures (chill injury)	Temperature sensors (minimum exposure)
Exposure to light	Photodetectors
Physical handling (e.g., bruising)	Shock sensors and vibration sensors

discoloration, heat, or gas production, etc.). The key environmental factors and the related sensor types are mentioned in Table 5.2.

The lifetime of perishable goods is influenced by environmental conditions such as temperature, relative humidity, and shock. Sensors can monitor these parameters and enhance logistic decision-making based on the actual quality level of goods. By using sensors, one does not need to perform a time-consuming task which might involve opening packaging or send off a sample for analysis. Instead, there will be potential for increased automation and reliable logging of data for traceability, quality assurance, and stock control. A further benefit is the ability to estimate a quality status of an item that might otherwise be difficult or impractical to determine (Bowman et al. 2009). With the sensor information, retailers can achieve higher profits, increased resource efficiency, higher quality, and a reduced amount of perished goods on the sales floor (Illic et al. 2008). Figure 5.10 shows how sensor-based approach is different from traditional- and labeling-based one in the management of stock rotation.

Fig. 5.10 Stock rotations based on different approaches

5.6   Wireless Sensor Networks and Internet of Things in Cold Chain

The Wireless Sensor Networks (WSN) is a type of network that emerges from the convergence of micro-electro-mechanical systems, wireless communications and digital electronics. It is a system which is capable of self-configuring, self-networking, self-diagnosing, and self-healing. It is considered as a very attractive solution for a wide range of environmental monitoring, distributed surveillance, health care, and control applications (Akyildiz et al. 2002). WSN is an ideal solution for physical and environmental monitoring. The main difference between a WSN and a RFID system is that RFID devices have no cooperative capabilities, while WSN allows different network topologies and multi-hop communication (Ruiz-Garcia et al. 2009). A typical WSN is shown in Fig. 5.11.

Carullo et al. (2009) introduced a WSN that was specifically designed to monitor temperature-sensitive products during their distribution with the aim of conforming to the cold chain assurance requirements. The proposed architecture is based on specifically designed nodes that are inserted into the products to identify their behavior under real operating conditions, e.g., during a typical distribution. Such product nodes communicate through a wireless channel with a base station, which collects and processes the data sent by all the nodes. A peculiarity of the product nodes is the low cost, which allows the information on the cold chain integrity to be provided to the final customer. The results that refer to the functional tests of the proposed system and to the experimental tests performed on a refrigerated vehicle during a distribution are reported.

Over the past few years, IoT has become one of the most important technologies of the twenty-first century. With the advancement of processing chips, wireless technology, and internet connectivity, the physical objects or "things" that are embedded with sensors, software, and other technologies can connect as a network and exchange the data with other devices and systems over the internet. The use

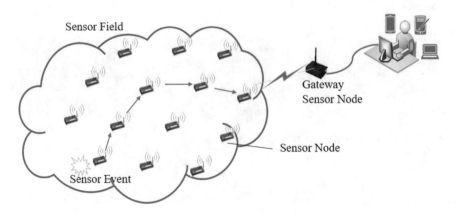

Fig. 5.11 Typical wireless sensor network

of artificial intelligence, machine learning, and embedding sensors to devices make a level of digital intelligence to devices and data collecting processes easier, more dynamic, and enabling them to communicate real-time data without interruption.

An IoT ecosystem consists of web-enabled smart devices that use embedded systems, such as processors, sensors, and communication hardware, to collect, send, and act on data they acquire from their environments. Wireless IoT monitoring enables to monitor the equipment in real time as well as receive immediate email or SMS alerts for abnormal conditions. This technology eliminates the possibility of human mistake, can store data for long periods of time, continuously monitors equipment, and promptly warns users when circumstances are found to be out of specification. Data are safely stored in the cloud and is easily accessible for regulatory purposes. Users receive alerts when battery power is low or connectivity issues arise, ensuring that no data are lost.

For example, IoT system for smart vending machine can check the operation status of the machine automatically without human intervention. Stock status, temperature status, selling status, and workable status will all be sent in real time to the monitoring center. Furthermore, when any aspect of the machine fails to satisfy the standards, monitoring centers will be notified. As a result, machine management will be more cost-effective, precise, and timely.

5.7   Integration of Tools and Technologies for Cold Chain

The tools and technologies that are used to monitor cold chain have different advantages in their own ways. In some situations, integrated solutions are necessary to ensure temperature integrity of the cold chain. Table 5.3 shows the common features and functions of different tools used for cold chain monitoring.

RFID can be integrated with sensors, thus are capable of measuring and acquiring data from the product behavior and the environment such as temperature, pressure, tampering, shock, humidity. Among environmental data, temperature data are critical since it is directly related to the product health and life cycle. A temperature-controlled supply chain is very important to ensure that the consumer receives high-quality and safe perishable products (e.g., food, vaccines, or pharmaceutical products). However, since RFID itself can only identify items and track the location of products, in order to be used for cold chain, temperature sensing technologies should be incorporated. A temperature-monitoring tag with RFID technology incorporates a sensor that captures the ambient temperature at specific intervals. Sensors are calibrated to detect high and/or low temperatures (Estrada-Flores and Tanner 2008).

Any temperature change triggers a visual indication on the tag and records the temperature, date, and time. RFID sensor tag is typically composed of:

a.   An antenna directly matched to the tag's frontend impedance to communicate with the reader.

Table 5.3 Cold chain monitoring tools

Tools	Key features	Functions
Thermometer/temperature probe Chart recorders TI TTI	– Instant visual feedback but no record for continuous monitoring – Charts are needed to replace – TI and TTI labels are non-reversible – Low cost and easy to use – With or without battery – No need technical ability	– Indicate current temperature (in thermometer) – Data shown in colored charts (in chart recorder) – Show temperature excursions by physical or color changes – Work with limited data, cannot archive
Temperature data loggers RFID WSN/IoT	– Automated and wireless – Monitor a wider area – Need technical ability – Power supply or Battery needed for continuous monitoring	– Assessing the data – triggering warning and alarms – Facilitating evaluation and acknowledgment – Performing regular reporting – Archiving the data – Automated, wireless solution

b. An analog RF front end that typically contains rectifier circuitry to convert RF power into DC, a clock, a modulator and a demodulator.

c. A logic part that is the translator between the front end and the sensor interface by coding, decoding, commanding, processing, and storing information. The logic implementation usually follows a defined standard and a certain associated protocol.

d. The signal interface that adapts the external signals (sensor reading, data logging, microcontrollers, display, keyboard…) to the standardized RFID tag (Pesonen et al. 2009).

RFID sensor tag can be illustrated as shown in Fig. 5.12.

Many companies today are adopting RFID and sensing technology for managing their cold chain. Unilever has completed a proof-of-concept test employing RFID tags and temperature sensors on cases of ice cream produced in the city of Veszprem, Hungary. During the trial, they found occasional breaks in the cold chain—particularly when goods were loaded on and off trucks (http://www.rfidjournal.com/articles/view?3863). Manor has been monitoring supermarket freezers and refrigerators using ultrahigh frequency (UHF) active RFID tags with built-in sensors to measure and log the temperatures of 1800 freezers and refrigerators every 10 min. The company is also developing plans to expand the application, so it can track refrigerated and frozen food as it moves from a supplier to Manor's distribution centers, then on to its retail outlets (http://www.rfidjournal.com/articles/view?3883).

The combination of RFID technology and time–temperature indicators (TTI) opens up the possibility of tracking the shelf life of chilled and frozen products remotely. In Europe, 16 million € project was partly funded by the European Commission and carried out by a consortium consisting of 26 partners from 13

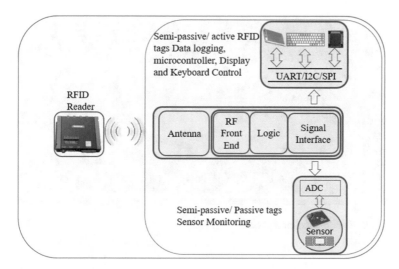

Fig. 5.12 Radio frequency identification sensor tag system architecture

different countries (http://www.chill-on.com). In the project, they also developed and validated cost-effective bio-sensing technologies for quantitative detection of microorganisms relevant to fish and poultry.

Alvin system has launched a cold chain solution with credit card-sized smart labels that combine 13.56 MHz RFID, temperature sensors, and mobile pocket PCs (www.alvinsystems.com). The RFID sensor tags monitor the condition of temperature-sensitive objects during transportation or storage—for quality assurance and enhanced cold chain operations.

RFID systems are typically used in track and trace applications, while sensor-based system or WSNs are deployed in monitoring applications. Many active, semi-passive, and semi-active tags have incorporated sensors into their design, allowing them to take sensor readings and transmit them to a reader. But, they are not quite sensor network nodes because they lack the capacity to communicate with one another through a cooperatively formed ad hoc network, but they are beyond simple RFID storage tags (Ho et al. 2005).

In most cases, no technology provides a complete solution for supply chain visibility. Although sensor-based system or WSN and RFID were originally invented with relatively different objectives (RFID for identification, while sensor-based system or WSN for sensing), they are complementary technologies and there exists a number of advantages in merging these two technologies. By integrating two technologies, a system can be equipped with sensing capabilities as well as tracking capabilities. WSNs offer a number of advantages over traditional RFID implementations such as multi-hop communication, sensing capabilities, and programmable sensor nodes (Liu et al. 2008).

Each technology solution has strength and weakness and will be applied for different requirements. Integrated use of two technologies together is effective in

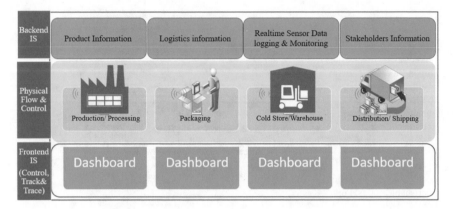

Fig. 5.13 Cold chain monitoring system

some applications. However, for example, integration of RFID and WSNs raises new challenges from the interference standpoint. There will be a lot of work to be done on how to reduce the interference in large RFID networks and WSNs since a large number of devices adversely increase the potential for interference. For WSN, a node's battery is not replaceable in real-time scenarios, so its energy is the most important system resource. An RFID system where data are collected at one or several centralized points is typically not energy-efficient. Therefore, energy efficiency will be a crucial problem when combining RFID and WSN and a good scheme to conserve energy is needed. The significant challenges will come with ubiquitous deployment of RFID and WSNs. Current RFID deployments have not yet included hundreds of readers in a single environment (Aung et al. 2013).

For complete view, Fig. 5.13 shows a cold chain system that consists of frontend information system (monitoring and control interface) which is supported by backend information system that monitors, collects, and uses data regarding physical flow and status of the products by using tools, embedded tags, and sensors from manufacturing stage to warehouse, transportation, and retail stage, ultimately the whole life cycle of a product.

References

Abad E, Palacio F, Nuin M, Zárate GD, Juarros A, Gómez JM, Marco S (2009) RFID smart tag for traceability and cold chain monitoring of foods: demonstration in an intercontinental fresh fish logistic chain. J Food Eng 93(4):394–399

Akyildiz IF, Su W, Sankarasubramaniam Y, Cayirci E (2002) A survey on sensor networks. IEEE Commun Mag 40(8)

Aung MM, Chang YS, Won JU (2013) Wireless sensor network and convergent technologies. Int J Adv Logist 1(2):46–69

Banks J, Hanny D, Pachano MA, Thompson LG (2007) RFID applied. Wiley

Bowman AP, Ng J, Harrison M, Lopez TS, Illic A (2009) Sensor based condition monitoring. Bridge project. www.bridge-project.eu

Carullo A, Corbellini S, Parvis M, Vallan A (2009) A wireless sensor network for cold-chain monitoring. IEEE Trans Instrum Measur 58(5):1405–1411

Estrada-Flores S, Tanner CJ (2008) RFID technologies for cold chain applications—review article. Bulletin 4

Ho L, Moh M, Walker Z, Hamada T, Su C (2005) A prototype on RFID and sensor networks for elder healthcare: progress report. In: Proceedings of the ACM SIGCOMM workshop on experimental approaches to wireless network design and analysis, August 22–26, Philadelphia, Pennsylvania, US

Ilic A, Staake T, Fleisch E (2008) The value of sensor information for the management of perishable goods—a simulation study. Bits-to-Energy Lab, Working Paper

Ilic A, Staake T, Fleishc E (2009) Using sensor information to reduce the carbon footprint of perishable goods. Pervas Comput IEEE 8(1):22–29

ISO/IEC 17364 (2013) ISO

Liu H, Bolic M, Nayak A, Stojmenovic I (2008) Taxonomy and challenges of the integration of RFID and wireless sensor networks. IEEE Network 22(6):26–35

Pesonen N, Jaakkola K, Lamy J, Nummila K, Marjonen J (2009) Smart RFID tags. In: Turcu C (ed) Development and implementation of RFID technology, InTech, pp 159–170

Ruiz-Garcia L, Lunadei L, Barreiro P, Robla JI (2009) A review of wireless sensor technologies and applications in agriculture and food industry: state of the art and current trends. Sensors 9:4728–4750

Sahin E, Badaï MZ, Dallery Y, Vaillant R (2007) Ensuring supply chain safety through time temperature integrators. Int J Logist Manag 18(1):102–124

Syntax Commerce (2007) Cold chain traceability. White paper

Chapter 6
Temperature Management in Cold Chain

The supply chain can be quite complex when dealing with food products. The limited lifetime and the deteriorating quality of perishable foods over time contribute substantially to the complexity of their management (Bowman et al. 2009). Successful Cold Chain Logistics calls for automated and efficient monitoring and control of all operations (Ruiz-Garcia and Lunadei 2010). Parties involved need better quality assurance methods to satisfy customer demands and to create a competitive point of difference. Cold chain typically focuses on temperature control and management in order to prevent the growth of microorganisms and deterioration of products during processing, storage, and distribution. It includes all segments in the food supply chain from the producer to the consumer. Therefore, it can be seen as a single entity since a breakdown in temperature control at any stage can impact the final quality of the product (Miles et al. 2008).

The preliminary tasks before storage are as important as the tasks after storage to maintain a sustainable and unbroken cold chain. In storing perishable products, one must consider product characteristics, especially temperature requirement. As explained in Sect. 4.2, every perishable products have different storage requirements (storage temperature, lifetime, humidity, etc.). A recent study shows temperature-controlled shipments from supplier to distribution center/store encounter the situation where the temperature rises or falls from the specified temperature account from 15 to 36% of trips (White 2007).

Unlike other products, fresh products have tough temperature requirement during their logistics processes because of its short shelf life and perishability. Temperature requirements vary among food items, whether frozen or chilled, and they even differ across types of foods. Even short exposures like a few hours to extreme hot or cold temperatures can cause a marked decrease in shelf life and loss of quality. Correct and careful temperature management throughout the supply chain is essential if quality of the product is to be assured (Jobling 2000).

© Springer Nature Switzerland AG 2023 93
M. M. Aung and Y. S. Chang, *Cold Chain Management*, Springer Series in Advanced Manufacturing, https://doi.org/10.1007/978-3-031-09567-2_6

Temperature is also a factor that can be easily and promptly controlled. Preservation of perishables' quality and safety can be only achieved when the product is maintained under its optimum temperature as soon as possible after harvest or production (Jedermann et al. 2007; Zhang et al. 2009). Chilled foods are easily temperature abused and temperature control and monitoring is an important factor in the control of safety and quality. There is also a need to maintain awareness for potential growth of microorganisms such as Listeria, Yersenia, and Aeromonas at chill temperatures (Martin and Ronan 2000). To overcome these time-related problems, reliable temperature management is essential to guarantee that the product arrives at its final destination in the best possible condition (Bowman et al. 2009).

Good temperature management is, in fact, the most important and simplest procedure for delaying the deterioration of food products. The storage at optimum temperature retards aging of fruits and vegetables, softening, and changes in texture and color, as well as slowing undesirable metabolic changes, moisture loss, and loss of edibility due to invasion by pathogens. Temperature control is thus essential and is an effective way of slowing bacterial growth, maintaining quality and minimizing spoilage. Conversely, high temperatures will cause an increase in the rate of bacterial growth, enzyme activity, and also other chemical reactions.

The optimization of temperature control in storage facilities such as refrigerated truck, warehouse, and cold store is crucial in today's operation of Cold Chain Logistics. The increased quantity and increased distance of food shipment are associated with logistics risks caused mainly by poor logistics technology and inefficient logistics management. These risks cause great damage in the agriculture sector as they lead to food loss, food contamination, spread of disease, and environmental damage. In order to counteract the logistics risks, more attention should be given to develop effective logistics technologies and efficient logistics management in the food supply system (Bosona 2013).

To achieve the monitoring and control of every link in a cold chain, real-time data should be communicated via data retrieval devices. The improved cold chain needs to be instrumented, interconnected, and intelligent. Cold chains need to be interconnected to customers, suppliers, and IT systems, as well as to products, trailers, and other smart objects that monitor the cold chain. Technologies such as sensors, radio frequency identification (RFID), wired, and wireless networks are potential components of a model to ensure an ongoing portable record of each product or its surroundings throughout its life cycle. Intelligent cold chains are those with advanced analytics and modeling based on food science and safety guidelines, which will assist managers with complex decisions in practical and efficient ways (Terreri 2009). Figure 6.1 shows typical cold chain management from raw material supplier to consumer and information to be managed throughout the chain.

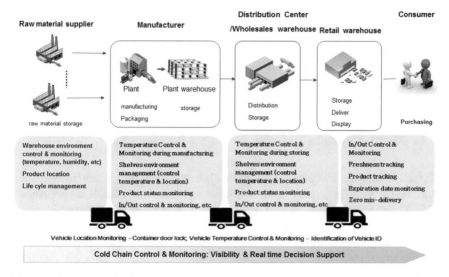

Fig. 6.1 Cold chain visibility and scope

6.1 Analysis on Product Characteristics of the Commodities

All cold chain products are not the same in terms of life span and storage conditions. The variety of the products and the diversified requirement in temperature or humidity contribute to the complexity of cold chain management. Before setting up a cold chain system, it is necessary for logistics managers to know the product characteristic (Zhang 2007). For fresh fruits and vegetables, it is important to check on product compatibility firstly. The ideal temperature often depends on the geographic origin of the product. Tropical fruits must be stored above 12 °C as they are not compatible to store with temperate one which can be stored at 0 °C (Jobling 2000). The knowledge on product perishability rate also helps to set priority for cooling and storage of the products.

Among perishable products, food has direct impact on the health of human and animal. Each food has different time limit depending on the product status and characteristics (e.g., opened, unopened, vacuum sealed, steak, chops, roast, etc.). The recommended time limits will help keep refrigerated food from spoiling or becoming dangerous to eat.

The refrigeration system should be designed to adjust and operate to a range of temperature and humidity conditions, depending on the compatibility group for storage of fruits and vegetables (NHB 2010). In general, mixed load or storage of perishables is not always recommended as there are concerns with the compatibility of the products such as temperature, relative humidity (RH), ethylene production/sensitivity, chilling or freezing sensitivity and off-odor or colors due to cross-contaminations. However, due to limited storage space and warehouse facility, it is quite common to keep different products in the same storage with inappropriate condition.

Furthermore, it is not easy to manage the requirement of products during the shipment by truck. Sometimes problems may not be visible but lead to a loss of quality for the consumer (Martin and Ronan 2000).

6.2 Multi-commodity Cold Storage Management

The single temperature system is dedicated to store food product with narrow temperature range; therefore, many more deliveries were needed to fulfill supply chain's demand for diverse products. Such systems were essentially inefficient and ineffective. The deliveries to the retail stores took place two or three times a week using the different temperature-controlled vehicle. In contrast, multi-temperature system has benefits that include improved vehicle utilization and improved service to retail. By using delivery vehicles with movable bulkheads and temperature-controlled evaporators, different temperature regimes could be set on one vehicle (Smith and Sparks 2004). However, the optimum storage conditions for each good have to be carefully evaluated. The desired transit temperature of products shipped together should be within reasonably close range. For example, mature green tomatoes requiring a transit temperature of 55 °F (13 °C) should not be shipped in combination with lettuce needing a transit temperature of 32 °F (0 °C) (Ashby 2006). In general, mixed load or storage of perishables is not always recommended in refrigerated storage as there are concerns with the compatibility of the products such as ethylene production/sensitivity, chilling or freezing sensitivity and off-odor or colors due to cross-contaminations. Sometimes problems may not be visible but lead to a loss of quality for the consumer (Martin and Ronan 2000).

There is a concern to decide whether the supply chain needs single or multi-commodity cold storage or refrigerated vehicle (Kuo and Chen 2010). Multi-commodity cold stores are provided with multiple chambers enabling them to store a wide range of fresh horticulture products together with respect to their storage compatibility requirements for temperature, RH, and atmosphere protection from odor and sensitivity to other gases like ethylene. It is a common practice for supermarket chains to deliver produce inside multi-compartment semitrailers. The different temperatures in each compartment are achieved by using distributed evaporator coils fed from a single condensing unit. Temperature distribution inside the compartment might be not equal, but any temperature configuration to adjust is possible. The design and control of refrigeration systems for multi-compartment vehicles are much more challenging than that for single compartment vehicles (Tassou et al. 2009).

It is important to make sure that all kind of perishable products with different temperature requirements can be maintained at the best quality condition, from the point of supply to the point of consumption, throughout the processes of storage and distribution. The freshness and safety of food have to be ensured in each stage in logistics service so as to maintain the value and quality to satisfy customers. All cold chain products are not the same in terms of life span and storage conditions. The variety of the products and the diversified requirement in temperature or humidity contribute to the complexity of cold chain management. Before setting up a cold chain system, it is necessary for logistics managers to know the product characteristic in the cold chain (Zhang 2007). At extreme temperatures, products are damaged. Some suffer chilling injury, some suffer damage at very high temperatures, and all products are damaged if they freeze at extreme low temperature (Jobling 2000). The more distant the real storage temperature of a product is of its ideal temperature, the greater the costs with the loss of quality of this product. It was verified that the cost relating to the loss of quality of products has great influence on the total storage cost (Borghi et al. 2009).

The facilities such as refrigerated containers, warehouses/cold stores, and cold display cabinets are needed to store and preserve perishable goods in good and quality state. However, due to variety of products and its different temperature requirements, the allocation of products to refrigerated facilities is complex. For example, the deposit of a supermarket has restrictions, such as its physical space, and it is impossible and impractical to create an environment with specific conditions to store each variety of a product (Borghi et al. 2009). Recently, different types of multi-temperature refrigerator have been developed and commercialized (Fig. 6.2). Among them, Kimchi refrigerator has a unique market position since it is designed specifically to meet the storage requirements of kimchi types and different fermentation processes.

Fig. 6.2 Multi-temperature refrigerator (Reprinted by permission of Hengel—http://www.hengel.com/)

Multi-commodity cold stores are provided with multiple chambers enabling them to store a wide range of fresh horticulture products together with respect to their storage compatibility requirements for temperature, RH, and atmosphere protection from odor and sensitivity to other gases like ethylene. The refrigeration system is designed to adjust and operate to a range of temperature and humidity conditions, depending on the compatibility group for storage of fruits and vegetables. Efficiency and performance in such cold stores are linked to appropriate storage systems which greatly optimize space, allow uniform air circulation through the produce, and meet the fundamental requirements of stock rotation which is time sensitive due to limited shelf life. This becomes more important for the cold stores being set up for the retail trade, export, and food processing industry. The design of the multi-commodity cold store facility and method of precooling depends on various factors like nature of product, category, and product type which determines the period of storage for example short-term storage (generally refer to as 7–10 days storage) or long/medium-term storage. Handling, stacking and storage methods, packaging, and frequency of entry and exists are also key deciding factors (NHB 2010).

Within the last decade, there has been a rapid growth in the use of multi-temperature trailers for food delivery operations, especially for fast food and independent grocery stores. Multi-temperature vehicles usually have three compartments separately controlled at 0 °F (−18 °C) or below for frozen foods, around 35 °F (2 °C) for chilled foods, and around 55 °F (13 °C) for chill-sensitive products (Ashby 2006).

The transports to large retailers are normally made in separate trucks for chilled, frozen, and ambient food, respectively, while express and home delivery or food transports to smaller retailers are usually made as combined transports from wholesale dealers. This means that food with different temperature demands is transported together in the same vehicle in order to increase the utilization of truck capacity and storage capacity while not delaying delivery. The advanced equipment such as cold boxes, cold cabins, and eutectic plates can facilitate agility and can improve the transportation performance better than a specialized refrigerated truck fleet (Olsson 2004; Kuo and Chen 2010). Factors specific to the design of multi-compartment refrigeration systems are the heat transfer between the compartments, product loading patterns, and door openings and method of temperature control (Tassou et al. 2009). Figure 6.3 shows different types of multi-temperature partitions for truck.

In order to control multi-temperature environment (such as Fig. 6.3), a system which can monitor multiple locations is important. Figure 6.4 shows a six-zone web-based temperature-monitoring system which can monitor multiple location and also send email alerts.

6.3 Optimal Target Temperature

Refrigerated storage is one of the most widely practiced methods of controlling microbial growth in perishable foods. Refrigeration maintains quality and prolongs shelf life by keeping the product temperature at the point, where metabolic and

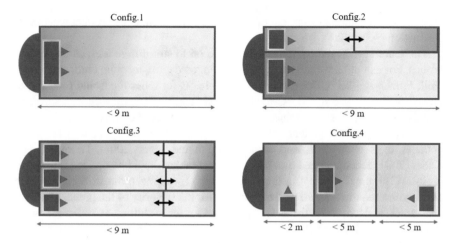

Fig. 6.3 Different types of multi-temperature partitions

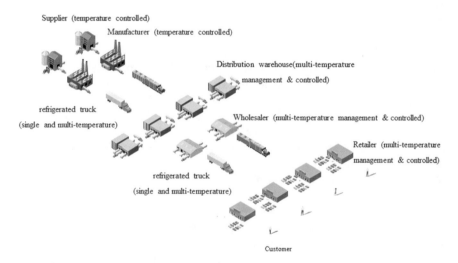

Fig. 6.4 Example of cold supply chain-based food logistics

microbial deterioration are minimized. However, refrigerated storage of perishable foods has been shown to be a potential risk factor for the development of microbial hazards leading to foodborne illness if proper temperature control is not conducted. The proper temperature control not only protects the safety but also maintains the quality of food (Jol et al. 2006). Maintaining the desired or ideal holding temperature is a major factor in protecting perishable foods against quality loss during storage and distribution. Quality loss is a function of both time and temperature abuse. Abuse is additive and, even for short periods of time during loading, transit, and unloading, may cause a considerable amount of quality loss by the time the product reaches its destination (Ashby 2006).

There are several temperature levels for food to suit different types of product groups. For example, we might identify frozen, cold chill, medium chill, and exotic chill: Frozen is $-25\,°C$ for ice cream and $-18\,°C$ for other foods and food ingredients; cold chill is 0 to $+1\,°C$ for fresh meat and poultry, most dairy and meat-based provisions, most vegetables, and some fruit; medium chill is $+5\,°C$ for some pastry-based products, butters, fats, and cheeses; exotic chill is $+10$ to $+15\,°C$ for potatoes, eggs, exotic fruit, and bananas (Smith and Sparks 2004). Thompson and Kader (2001) recommended that fruits and vegetables should be divided into three: (1) 0–2 °C group; (2) 7–10 °C group; and 13–18 °C group according to their optimum temperature requirements. The first group is for the majority of the green, non-fruit vegetables, and temperate fruits. Groups 2 and 3 refer to chilling-sensitive products. Publications such as the IIR Recommendations for the Chilled Storage of Perishable Produce (2000), Recommendations for the Processing and Handling of Frozen Foods (2006), and USDA Agriculture Handbook No. 66 (2016) provide data on recommended temperature and the storage life of many foods.

A temperature-controlled supply chain includes all storage and transport facilities necessary to ship a temperature-sensitive product from manufacturer to end user and is characterized by the demand of being frozen, chilled, or ambient (Taylor 2001). The most sensitive type of food in this respect is chilled food, while frozen food is less sensitive due to the lower temperatures and ambient food is less sensitive due to other processing or packaging techniques (Olsson 2004). Therefore, the refrigerated product including produce has the greatest losses, while frozen is the least (Labuza et al. 2003). Moreover, if the temperature of some chilly foods exceeds specific limits, a great decrease of the quality, along with the increase of the risk of food poisoning, can take place. The limits can be quite strict for chilled products with storage temperature near to 0 °C, where the rise in temperature of just a few degrees can cause microbial growth (Carullo et al. 2009).

If a food supply chain is dedicated to a narrow range of products, the temperature will be set at the level for that product group. However, if a food supply chain is handling a broad range of products, then an optimum temperature or a limited range of different temperature settings is used. Failure to maintain appropriate temperature regimes throughout a product's life can shorten the life of that product or adversely affect its quality or fitness for consumption (Smith and Sparks 2004). There is an optimal storage temperature range for all products. The temperature of refrigerated room for the product should be set at optimal target value because the temperature fluctuations from optimal value can deteriorate the quality of the product. For a range of optimal temperature, mean value represents target temperature value for that product. It is found that the ideal temperature often depends on the geographic origin of the product. Tropical plants have evolved in warmer climates and therefore cannot tolerate low temperatures during storage. Plants from tropical origins must be stored above 12 °C. This contrasts with plants which have evolved in temperate, cooler climates which can be stored at 0 °C (Jobling 2000).

6.4 Methods to Define Optimal Target Temperature

Typically, every produce which needs refrigeration has designated optimal temperature range or specific temperature point. The target optimal temperature which is destined for single product cannot cover for others if we store all together in the same cold storage area. There are varieties of products which demand chilled temperature condition, and the temperature management for these products is much more complex than frozen products. Aung and Chang (2014) proposed the methods to define optimal target temperature for fruits that have different storage temperatures. The sample dataset includes recommended storage temperature and related information about temperate, tropical, and subtropical fruits (Table 6.1). The basic principle is that the range of their recommended temperatures (lower and upper level) can be considered as coordinate points (x, y) in a plane. In multi-commodity cold storage environment, it is important to investigate the appropriate methods to define the optimal target temperature value that has minimal impact and works for all commodities. However, no standard method has been defined yet for such a selection of optimal target temperature other than choosing an estimated optimal value. In their

Table 6.1 Sample dataset for tropical and subtropical fruits

Commodity	Min_ temp (°C)	Max_ temp (°C)	Units	Unit_Price ($)	Total Amt. ($)	Vol. (cubic feet)	Total Vol. (cubic feet)	Shelf_Life (days)
	x	y	n	p	$P (p * n)$	v	$V (v * n)$	l
A (Banana)	13	16	10	5	50	2	20	12
B (Cranberries)	3	6	15	10	150	1	15	10
C (Grapefruits)	13	16	14	15	210	1	14	9
D (Tangerines)	4	7	13	7	91	2	26	5
E (Guavas)	7	10	12	12	144	2	24	4
F (Lemons)	11	13	11	5	55	1	11	7
G (Lychees)	4	7	10	25	250	3	30	9
H (Mangoes)	10	13	9	8	72	2	18	8
I (Melons)	10	13	8	20	160	3	24	2
J (Oranges)	4	7	7	15	105	1	7	1
K (Persimmons)	0	2	6	23	138	1	6	11
L (Strawberries)	0	2	5	7	35	2	10	3
M (Pomegranates)	5	10	4	8	32	3	12	8
N (Avocados)	3	7	3	10	30	2	6	4
O (Apples)	0	4	2	11	22	4	8	6
P (Blackberries)	0	1	2	8	16	2	4	3
Q (Tomato)	13	21	4	6	24	4	16	4

study, the three methods centroid method, weighted centroid method and clustering method are proposed and evaluated.

6.4.1 Centroid Method

The concept of centroid is the multivariate equivalent of the mean. The conventional and simplest method to find the mean is a coarse-grained range-free centroid localization algorithm which needs only a minimum of computations (Bulusu et al. 2000). This method finds the centroid of any set of points on the x–y plane.

For a number of points (n), i.e., (x_1, y_1), (x_2, y_2), ... , (x_n, y_n),

$$\text{the centroid location}\quad (x^*, y^*) = \left[\frac{x_1 + \cdots + x_n}{n}, \frac{y_1 + \cdots + y_n}{n} \right] \quad (6.1)$$

Using the data in Table 6.1, the centroid location (x^*, y^*) can be found at (5.88, 9.12).

The distinguished advantage of this centroid localization method is its simplicity and ease of implementation. But, this centroid location estimation formula considers finding centroid location based on coordinate (x, y) information excluding other factors. However, this is a useful method applied as a part of k-means algorithms to find the centroid value of clusters.

6.4.2 Weighted Centroid Method

The improved and alternative approach to determine optimal target temperature is weighted centroid or center of gravity (CG) method which is commonly used in facility location problems. The gravity problem corresponds to the case of an objective equal to the square of the Euclidean distance. Assume that the existing facilities are located at points (a_1, b_1), (a_2, b_2), ... , (a_n, b_n). The weight (w) is included to allow different transportation rates/cost between new point and existing points where the cost is proportional to the square of the distance traveled (Nahmias 2005). The objective is to find values of (x, y) to minimize

$$f(x, y) = \sum_{i=1}^{n} w_i [(x - a_i)^2 + (y - b_i)^2] \quad (6.2)$$

The optimal values of (x, y) are easily determined by differentiation. The final optimal solution with respect to x and y is

$$x^* = \frac{\sum_{i=1}^{n} w_i a_i}{\sum_{i=1}^{n} w_i}, \quad y^* = \frac{\sum_{i=1}^{n} w_i b_i}{\sum_{i=1}^{n} w_i} \tag{6.3}$$

Therefore, for Table 6.1, the optimal value of (x^*, y^*) is at (6.32, 9.27).

This is an approach that can be easily adopted to find optimal target temperature by using data points which represent storage temperature and total value of the products. Data points refer to the storage temperature of the product, and the value can be considered as the weight (w). However, the application of the above approaches shows that the result does not fit well with the objectives. It is found that the optimal target temperature can effectively work for a limited range of products. Since the temperature range for chilled products covers from 0 to 21 °C, it is difficult to maintain the total quality of products by setting single optimal temperature for cold storage. The best way to solve this problem is to divide and conquer temperature ranges of the products. Clustering, one of the data mining techniques, is found to be useful in making a set of clusters which has similar pattern or characteristics. It is apparent that the improved, more balanced, and integrated approach is needed in solving the problem.

6.4.3   Clustering-Based Method

The process of grouping a set of physical or abstract objects into classes of similar objects is called clustering. A cluster is a collection of data objects that are similar to one another within the same cluster and are dissimilar to the objects in other clusters. A cluster of data objects can be treated collectively as one group in many applications. Contributing areas of research include data mining, statistics, machine learning, spatial database technology, biology, and marketing. As a branch of statistics, cluster analysis has been studied extensively for many years, focusing mainly on distance-based cluster analysis. In machine learning, clustering is an example of unsupervised learning (Han and Kamber 2006). The k-means algorithm is a simple yet effective statistical clustering technique. In this algorithm, firstly, it randomly selects k of the objects, each of which initially represents a cluster mean or center. For each of the remaining objects, an object is assigned to the cluster to which it is the most similar, based on the distance between the object and the cluster mean (i.e., Euclidean distance). It then computes the new mean for each cluster. This process iterates until the criterion function converges (Roiger and Geatz 2003).

According to the data mentioned in Table 6.1, it is inappropriate to keep all chilled products in single area due to wide temperature coverage of chilled products (0–21 °C). Two or more refrigerated storages with different temperature zones are necessary to fulfill the temperature requirement of these products. One of the algorithms that suits for that purpose is clustering algorithm. The number of cluster through k-means represents the refrigerated room needed for products. We consider using k-means clustering algorithm on the sample dataset of fruits. However, one of its drawbacks is the requirement for the number of clusters, k, to be specified before

the algorithm is applied (Pham et al. 2005). Typically, a refrigerated container or
warehouse uses two–three rooms for chilled produce, but the case that needs four
rooms is also included in the simulation. Therefore, k values of 2, 3, and 4 are applied
to decide the right number of rooms or temperature zones needed for storage.

The k-means algorithm can be used for grouping any set of objects whose simi-
larity measure can be defined numerically. For example, a set of records of a
relational-database table can be divided into clusters based on any numerical field
of the table. Traditionally, forming clusters using k-means usually based on the
Euclidean distance of data points, but if those data points have some related data, we
can also consider the distance measure on the value of these related data fields.

Figure 6.5 shows the steps to find optimal target temperature:

- Step 1. k-means clustering is applied on the sample dataset with $k = 2$, 3, and 4.
  Based on k values, the different clusters are formed.
- Step 2. Again, k-means clustering is applied to each cluster with $k = 2$ for two
  times, but this time, it will be based on two criteria: the total value of produce and
  the shelf life. Depending on each criterion, different number of subclusters will
  be formed.
- Step 3. From each pair of value- or shelf life-based subclusters, the subclusters
  that have maximum total average value and minimum total average shelf life will
  be retrieved.
- Step 4. The centroid value is calculated from the resulting two subclusters to get
  final target temperature range.

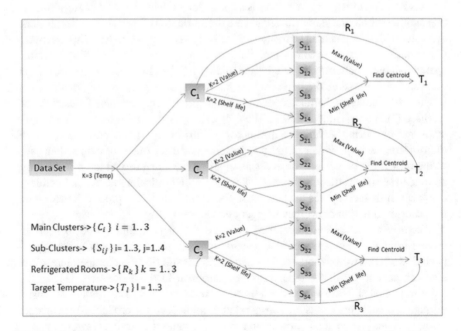

Fig. 6.5 Procedure flow to get optimal target temperature using k-means clustering ($k = 3$)

This approach is a little different from CG method as adjusting the optimal target temperature is adjusted using both total value and the shelf life not to lose quality quickly for products that has a few days of remaining shelf life.

Through clustering, the chilled products are classified into a set of clusters that have different but manageable temperature zones. Then the optimal target temperature is decided based on three methods: centroid method that is based on only the value of temperature, CG method that accounts weight (i.e., cost/value), and the balanced method that considers not only weight but also remaining shelf life of the products.

Sample dataset Table 6.1 includes the products that have different storage temperatures ranging from 0 to 21 °C. The result of optimal target temperatures for different groups (i.e., 2, 3, and 4 clusters) is illustrated in Table 6.2.

The current typical multi-temperature vehicles have three compartments: one for frozen; two for chilled; and chilled-sensitive foods. The specific optimal temperature of the products should be as close as target optimal temperature to reduce product deterioration of the group. In that sense, four-clusters ($k = 4$) solution (Fig. 6.8) is the best among three solutions as it has temperature zones in small range; (0–4 °C), (3–10 °C), (7–13 °C), and (13–21 °C). However, dividing many rooms has concerns over cost, poor utilization of space, and energy consumption than using small number of rooms. Three clusters ($k = 3$) solution (Fig. 6.7) is found to be an appropriate choice as its temperature zones cover for (0–4 °C), (3–10 °C), and (10–21 °C). If refrigerated space is limited, two clusters ($k = 2$) solution (Fig. 6.6) is found suitable as the first two clusters can be combined into one like (0–10 °C) and (10–21 °C). This concept of combination is found to agree the proposal of (Thompson and Kadar 2001) to combine two groups (0–2 °C) and (7–10 °C) for limited space problem.

Table 6.2 Optimal target temperatures for different scenarios

Methods	Cluster 1	Cluster 2	Cluster 3	Cluster 4
	(x, y)	(x, y)	(x, y)	(x, y)
Centroid	(11.67, 15.33)	(2.73, 5.73)		
CG	(11.59, 14.7)	(3.45, 5.53)	–	–
k-means ($k = 2$)	(11.5, 15.75)	(3.395, 5.86)	–	–
Centroid	(11.67, 15.33)	(4.28, 7.71)	(0, 2.25)	–
CG	(11.59, 14.7)	(4.35, 6.42)	(0, 2.13)	–
k-means ($k = 3$)	(11.5, 15.75)	(4.59, 7.71)	(0, 2.17)	–
Centroid	(13, 17.66)	(9.5, 12.25)	(3.83, 7.33)	(0, 2.25)
CG	(13, 16.42)	(9.12, 12)	(3.78, 5.64)	(0, 2.13)
k-means ($k = 4$)	(13, 18.5)	(8.5, 11)	(3.59, 6.75)	(0, 2.17)

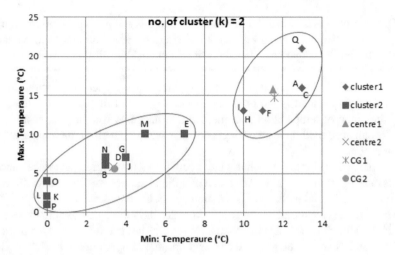

Fig. 6.6   Clustering the dataset using k-means ($k = 2$)

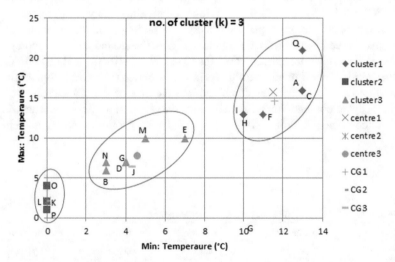

Fig. 6.7   Clustering the dataset using k-means ($k = 3$)

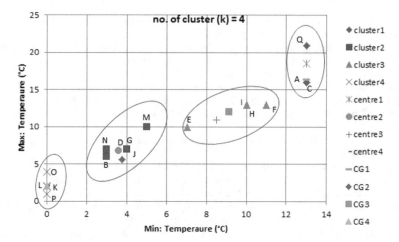

Fig. 6.8 Clustering the dataset using k-means ($k = 4$)

References

Ashby BH (2006) Protecting perishable foods during transport by truck. Handbook no. 669, United States Department of Agriculture

Aung MM, Chang YS (2014) Temperature management for the quality assurance of a perishable food supply chain. Food Control 40:198–207

Borghi DF, Guirardello R, Filho LC (2009) Storage logistics of fruits and vegetables: effects of temperature, web papers

Bosona T (2013) Logistics risks in the food supply chains. Forum for Agricultural Risk Management

Bowman AP, Ng J, Harrison M, Lopez TS, Illic A (2009) Sensor based condition monitoring. Bridge Project. www.bridge-project.eu

Bulusu N, Heidemann J, Estrin D (2000) GPS-less low cost outdoor localization for very small devices. IEEE Pers Commun Mag Newblock 7(5):28–34

Carullo A, Corbellini S, Parvis M, Reyneri L, Vallan A (2009) A wireless sensor network for cold-chain monitoring. IEEE Trans Instrum Measur Vancouver, Canada 58(5):1405–1411

Han J, Kamber M (2006) Data mining concepts and techniques, 2nd edn. Morgan Kauffmann Publishers

Jedermann R, Edmond JP, Lang W (2007) Shelf life prediction by intelligent RFID. In: Haasis HD, Kreowski HJ, Scholz-Reiter B (eds) Dynamics in logistics—first international conference, LDIC 2007 Bremen, Germany. Springer, Berlin, pp 231–238

Jobling J (2000) Temperature management is essential for maintaining produce quality. Good Fruit Veg Mag 10(10):30–31

Jol S, Kassianenko A, Wszol K, Oggel J (2006) Process control issues in time and temperature abuse of refrigerated foods. Food Safet Mag (Dec 05/Jan 06)

Kuo JC, Chen MC (2010) Developing an advanced multi-temperature joint distribution system for the food cold chain. Food Control 21(4):559–566

Labuza T, Belina D, Diez F (2003) Food safety management in the cold chain through "expiration dating". In: 1st international workshop cold chain management. University of Bonn, Germany

Martin G, Ronan G (2000) Managing the cold chain for quality and safety. Flair-flow Europe technical manual 378A/00

Miles SB, Sarma SE, Williams JR (2008) RFID technology and applications. Cambridge University Press

Nahmias S (2005) Facilities layout and location. In: Production and operation analysis, 5th edn. McGraw-Hill International Edition, pp 586–593

NHB (2010) Technical standards and protocol for the cold chain in India. In: Cold storage for fresh horticulture produce requiring pre-cooling before storage (technical standards number NHB-CS-Type 02–2010), National Horticulture Board (NHB), Govt. of India

Olsson A (2004) Temperature controlled supply chains call for improved knowledge and shared responsibility. In: Aronsson H (eds) Conference proceedings NOFOMA2004, Linköping, pp 569–82. Retrieved from https://lucris.lub.lu.se/ws/files/5817800/625938.pdf. Accessed 15 May 2020

Pham DT, Dimov SS, Nguyen CD (2005) Selection of K in K-means clustering, IMechE Part C. J Mechan Eng Sci 219:103–119

Roiger RJ, Geatz MW (2003) Data mining a tutorial based primer. Addison-Wisley

Ruiz-Garcia L, Lunadei L (2010) Monitoring cold chain logistics by means of RFID. In: Turcu C (ed) Sustainable radio frequency identification solutions, Croatia, Intech, pp 37–50

Smith D, Sparks L (2004) Temperature controlled supply chain. In: Bourlakis MA, Weighman PWH (eds) Food supply chain management. Blackwell Publishing, pp 179–198

Tassou SA, De-Lille G, Ge YT (2009) Food transport refrigeration—approaches to reduce energy consumption and environmental impacts of road transport. Appl Therm Eng 29:1467–1477

Taylor J (2001) Recommendations on the control and monitoring of storage and transportation temperatures of medicinal products. Pharm J 267

Terreri A (2009) Cold chain management: monitoring each link in the chain, food logistics

Thompson JF, Kader AA (2001) Wholesale distribution centre storage. Perishable Handl Q 107

White J (2007) How cold was it? Know the whole story. Frozen Food Age 56(3):38–40

Zhang L (2007) Cold chain management. In: Cranfield Centre for Logistics & Supply Chain Management, Cranfield University

Zhang J, Liu L, Mu W, Moga LM, Zhang X (2009) Development of temperature-managed traceability system for frozen and chilled food during storage and transportation. J Food Agric Environ 7(3&4):28–31

Chapter 7
Quality Assessment in Cold Chain

7.1 Quality Assessment Using Wireless Sensors

There is frequent loss of quality and value when goods are stored and transported. Quality needs to be assured that the products are kept at the prime conditions along the supply chain. The deterioration of perishable foods can lead to a decrease in the aesthetic appeal as well as a reduction in nutritional value. The degradation of quality in some food is readily visible by changes of texture or colors, but there are some situations where the degradation might not be so readily visible. However, the visual judgment on current quality is very subjective and not reliable in practice. Therefore, the food industry needs appropriate methods to overcome these challenges.

Real-time monitoring and quality assessment are found essential for safety, quality, and traceability of product, so visibility along the chain and confidence of the customers could be achieved. To assess the quality, the monitoring and collection of quality data through indicators using sensors are needed to calculate and judge the actual quality of the product. Wireless sensor network (WSN) is an efficient tool to monitor and controls the cold chain with respect to the loss of quality in perishable products during transportation and storage.

In the case of perishable goods, there is generally a need to monitor goods continuously, since there could be various changes which impact on the quality: a degradation of quality, creation of a hazard, reduction of value or reduced shelf life, etc. WSNs are ideally suited to monitor goods that could be degraded in quality over time. They are really powerful not only to measure direct environmental conditions (i.e., temperature, humidity) before (and during) deterioration but also to indicate the extent of progress or rate of degradation (i.e., dis-coloration, heat, or gas production, etc.). While the human senses have only a limited capability to assess the intrinsic product properties, modern sensor technologies can help to provide the required information.

© Springer Nature Switzerland AG 2023

M. M. Aung and Y. S. Chang, *Cold Chain Management*, Springer Series in Advanced Manufacturing, https://doi.org/10.1007/978-3-031-09567-2_7

7.2 Respiratory Metabolism

Fresh fruit and vegetables are living products. After harvest, they continue the process of respiration which produces carbon dioxide (CO_2), water, and heat. In climacteric fruit, the respiration typically rises very rapidly during ripening and then decreases as the fruit age after its maturity. The rate of deterioration of the produce is largely determined by the rate of respiration which is temperature dependent. Slow respiration rate can minimize product deterioration, but respiration can never be completely stopped. Produce which is kept cool will have a low rate of respiration with limited heat production and low rate of deterioration. Different products have different rates of respiration. Those with higher rates are more highly perishable, and temperature control is very critical for these products. Ethylene is produced by many plant products and can trigger ripening and deterioration in some products. Keeping products cool reduces the production of ethylene. Also, cooled products are less sensitive to ethylene (SARDI 2006). The climacteric and non-climacteric pattern in respiration of fruits is represented in Fig. 7.1.

In general, the storage life of commodities varies inversely with the rate of respiration. This is because respiration supplies compounds that determine the rate of metabolic processes directly related to quality parameters, e.g., firmness, sugar content, aroma, flavor, etc. Measurements of respiration provide an easy, non-destructive means of monitoring the metabolic and physiological state of tissues. Aerobic respiration can be seen in following Eq. (7.1) (Salveit 2004):

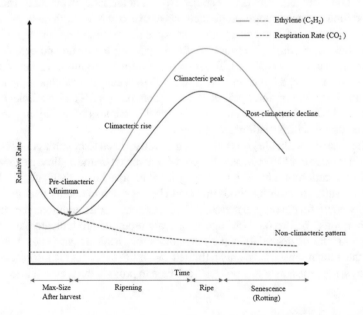

Fig. 7.1 Climacteric and non-climacteric pattern of respiration and ethylene emission in ripening fruit

$$\text{Glucose} + \text{Oxygen} \rightarrow \text{Carbon Dioxide} + \text{Water} + \text{Energy}$$
$$C_6H_{12}O_6 + 6O_2 \rightarrow 6CO_2 + 6H_2O + 686\,\text{kcal/mole} \qquad (7.1)$$

7.3   Quality Assessment Methodologies

The method to calculate freshness is based on the metabolic changes of perishable foods especially in harvested commodities. Freshness is considered alternatively as quality of the product (Kim et al. 2015). Initially, the freshness gauge (FG) of a new product item after harvest is assumed to be 100% Then, FG value can be changed whenever the temperature fluctuations happen. The amount of degradation depends on how much real (relative) temperature value is different from ideal (absolute) optimal one. FG is calculated based on product shelf life and current temperature value received from temperature sensor. The value of FG (Eq. 7.2) can be calculated as follows:

$$FG = \text{Previous FG} - \frac{(\text{New Log Time} - \text{Previous Log Time})}{(\text{Shelf Life} * \text{Weight of Shelf Life})} * 100 \qquad (7.2)$$

In this section, we discuss two methods which support required weight of shelf life value for FG formula. The first one is based on continuous monitoring on metabolic changes of fruit, and the second one relates to finding Euclidean distance (cost) of absolute and relative temperatures. In addition, a methodology based on visual quality assessment is added as a helpful tool for quality evaluation.

7.3.1   Respiratory Rate (Products' Metabolism)-Based Measurement

Temperature data are periodically recorded in short time interval and logged into database. FG value is adjusted according to the values of weight. Weight of shelf life calculation used temperature data which are received by monitoring system. Q_{10} temperature coefficient was adopted as an indicator to evaluate the quality changes during storage of perishable products (Aung et al. 2012). Q_{10} refers to the rate of change of a biological or chemical system as a consequence of increasing the temperature by 10 °C. The Q_{10} can be calculated by dividing the reaction rate at a higher temperature by the rate at a 10 °C lower temperature, i.e., $Q_{10} = R2/R1$ (Salveit 2004).

The respiration rate is typically measured by emission of CO_2 as a mechanism of respiration and is affected by temperature changes. However, the respiration rate does not follow ideal behavior, and the Q_{10} can vary considerably with temperature. To measure precisely the changing respiration rate and quality degradation, the measurement should be in smaller gap such as 2 °C or 3 °C. Using this data of measurement for specific commodity, the relative rate of deterioration and relative shelf life can be defined.

It is found that Q_{10} value can be applied in accelerated shelf life studies. Accelerated aging test parameters are based on Q_{10} thermodynamic temperature coefficient. The Arrhenius Reaction Rate theory states that a rise in temperature of 10 °C will roughly double the rate of chemical reaction (Bowman et al. 2009). The accelerated aging rate is given by the following equation (Shema 2010):

$$\text{Accelerated Aging Rate (AAR)} = Q_{10}^{((Te/Ta)/10)} \tag{7.3}$$

where

T_a ambient temperature;
T_e elevated temperature;
Q_{10} reaction rate (for 10 °C gap).

Then, the accelerated aging time duration (AATD) is given by the equation:

$$\text{AATD} = \text{Desired Real Time (Absolute Shelf Life)}/\text{AAR} \tag{7.4}$$

7.3.2 Euclidean Distance-Based Quality Measurement

The second approach to monitor quality is based on weight centroid (CG) method that uses center of gravity to find optimal single location such as terminal, warehouse, or retailer service points. The approach is simple, since transportation cost between the source and the material is only considered as location factor. The model is based on minimizing the total cost of transportation, i.e.,

$$\text{Min TC} = \sum_{i=1}^{n} Q_i C_i d_i \tag{7.5}$$

where

Q_i quantity of material shipped per unit period;
C_i cost of transporting a unit per unit distance;
d_i squared Euclidean distance between source point and demand point;
TC   total transportation cost (Raju 2008).

According to Eq. 7.2, $w_i = Q_i * C_i$ and $d_i = (x - a_i)^2 + (y - b_i)^2$.

At this point, the distance (d_i) in Eq. 7.2 refers to squared Euclidean distance and does not represent the straight line distance between two points. In our scenario, the following Euclidean distance (Eq. 7.6) is more preferable to use as it measures the straight line distance. Therefore,

$$d_i = \sqrt{(x - a_i)^2 + (y - b_i)^2} \tag{7.6}$$

Then, we can modify the equation as

$$\text{Min TC} = \sum_{i=1}^{n} w_i \sqrt{(x - a_i)^2 + (y - b_i)^2} \tag{7.7}$$

After clustering the dataset given in Table 6.1, we will get two or more clusters that represent different temperature zone (Z) where a group of products (C) can be stored; each one has its specified optimal temperature range (i.e., (a, b)) and associated cost (i.e., w). We can get the final absolute optimal target temperature (AOTT) after running the procedure mentioned in Fig. 7.2. However, optimal target temperature is changing dynamically in real time, and mostly relative optimal target temperature (ROTT) can be achieved. If we define AOTT as optimal state, there will be one or more ROTT states as storage temperature value might change over time. This is shown in Fig. 7.2.

Therefore, total transportation cost for each member of cluster can be calculated by using AOTT and dynamically changing ROTT. The difference between AOTT and ROTT will indicate how ROTT deviates from AOTT and also the associated cost obtained that related to the loss of quality. This could be applied to estimate how quality changes happen over temperature rise. It can be written as

$$\Delta TC = |TC \text{ for AOTT} - TC \text{ for ROTT}| \tag{7.8}$$

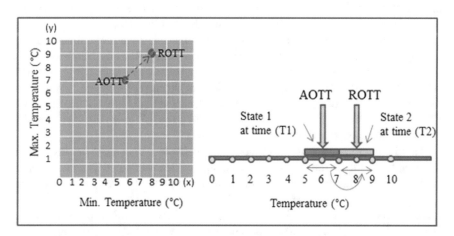

Fig. 7.2 Graphical view on changes of optimal target temperature over time (Aung 2013)

$$\Delta TC = \left| \sum_{i \in Z} \sum_{j \in C} w_{ij} \sqrt{\left(x^* - a_{ij} \right)^2 + \left(y^* - b_{ij} \right)^2} \right.$$

$$\left. - \sum_{i \in Z} \sum_{j \in C} w_{ij} \sqrt{\left(x - a_{ij} \right)^2 + \left(y - b_{ij} \right)^2} \right| \qquad (7.9)$$

where

ΔTC the difference in total cost between AOTT and ROTT;
x^* absolute min: temperature value;
y^* absolute max: temperature value;
x relative min: temperature value;
y relative max: temperature value.

The following restrictions are applied in this formulation: Each product can be assigned to only one specific zone or refrigerated room; the storage capacity of the zone cannot be exceeded; the time that product i stays in zone j depends on remaining shelf life value.

7.3.3   Visual Quality Assessment Methodology

Temperature is the key to keep the quality and integrity of product after harvest in cold chain. The effect of temperature on climacteric fruits is demonstrated by climacteric characters such as the changes in respiration rate, color, texture, and flavor. The climacteric character represents an important determinant of the ripeness, maturity, and storability. Therefore, these fruits should be maintained under properly controlled temperature not to lose the quality throughout the supply chain.

Color is one of the most important quality attributes in postharvest food handling and processing, and it influences consumer's choice and preferences. Banana industry today uses color charts to judge the skin color of banana as shown in Fig. 7.3. The color of the banana skin is used as an indicator for ripeness or spoilage. A peel color index (scale of 1–7) of banana is divided into seven levels: (1) green, (2) green-trace, (3) more green than yellow, (4) more yellow than green, (5) green tip, (6) all yellow, and (7) yellow-flecked with brown (Kerbel 2004).

These color charts are useful to judge the peel colors of banana, but inspector's decision on quality is rather visual (i.e., subjective) measurement. The instrumental (objective) methods are needed to determine more precisely over the maturity and quality of wide range of food products. There are some devices available such as colorimeter or chromameters for measuring reflected and transmitted color of objects. In Plainsirichai and Turner (2010), banana fruits were exposed to temperature of 20–30 °C to measure change in green color in the banana peel. The greenness of the peel of ripening banana fruit, expressed as a^* values (L^*, a^*, b^* color system), is

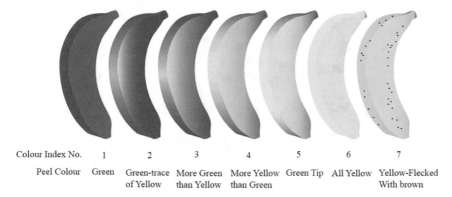

Colour Index No.	1	2	3	4	5	6	7
Peel Colour	Green	Green-trace of Yellow	More Green than Yellow	More Yellow than Green	Green Tip	All Yellow	Yellow-Flecked With brown

Fig. 7.3 Sample of banana color chart

measured with a chromameter. The more negative the value of a^*, the greener the appearance of the fruit is.

Alternatively, digital imaging method is found to be useful in fresh produce industry to estimate the ripening stages of bananas more accurately (Ji et al. 2012). A method to differentiate ripening skin colors of banana by using a software tool (i.e., Histogram of Adobe's Photoshop CS software) is found useful here. A histogram illustrates how pixels in an image are distributed by graphing the number of pixels at each color intensity level. The histogram panel offers many options for viewing tonal and color information about an image. The relative color intensity data can be collected after measurement of consecutive images at different stage of supply chain. Then, the level of the ripeness could be easily determined by histogram matching of the measured and standard ripeness stages.

7.3.4   Temperature Management and Quality Assessment

In Chaps. 6 and 7, three main methods were discussed, i.e., clustering-based temperature management for refrigerated storage, FG calculation to determine current freshness, and assessment of quality through monitoring quality parameters. In Fig. 7.4, the way how these methods relate to cold chain management is illustrated. The area where the methods applied relates to refrigerated storage, distribution, and transport process in cold chain management. The first method includes two steps, i.e., clustering the products and defining optimal target temperatures. Actually, these are preliminary tasks to do for refrigerated storage management to achieve optimal setting and not to loss quality. The second method relates to the calculation of freshness to assess the quality changes of the product based on product deterioration rate which can be calculated via sensors data collected. The third method supports freshness calculation by collecting continuous data over deterioration of product through

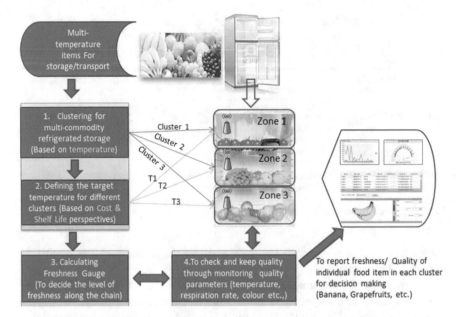

Fig. 7.4 Quality control and assurance methods applied in cold chain management

monitoring of quality parameters by using sensors and reports to user for decision-making. To assess the quality, we consider two types of measurement using sensors. The first sensor group which includes temperature, humidity, etc., is intended for direct sensing to the storage environment and second group which include CO_2, ethylene, etc., are used to estimate quality degradation (rate) via metabolic changes which relate to product deterioration process. There are three typical dimensions commonly employed (i.e., "How many?", "How high?", and "How long?") in the interpretation of sensor data over the breach of optimal temperature threshold value (Bowman et al. 2009). The FG calculation considers all these dimensions, so estimation for quality status of products in real-time monitoring environment could be easily achieved. There are other alternative methods suggested for verification of quality: Euclidean distance-based quality measurement and visual quality assessment method.

References

Aung MM (2013) Implementation of quality control and assurance methods for a climacteric fruit in cold chain, PhD thesis, Korea Aerospace University, Republic of Korea

Aung MM, Chang YS, Kim WR (2012) Quality monitoring and dynamic pricing in cold chain management. Proc World Acad Sci Eng Technol 62:435–439

Bowman AP, Ng J, Harrison M, Lopez TS, Illic A (2009) Sensor based condition monitoring. Bridge Project. www.bridge-project.eu

Ji W, Koutsidis G, Luo R, Hutchings J, Akhtar M, Megias F, Butterworth MA (2012) Digital imaging method for measuring banana ripeness. Color research & application, Wiley Online Library

Kerbel E (2004) Banana and plantain, the commercial storage of fruits, vegetables, and florist and nursery stocks. Agriculture handbook number 66, USDA

Kim WR, Aung MM, Chang YS, Makatsoris C (2015) Freshness Gauge based cold storage management: a method for adjusting temperature and humidity levels for food quality. Food Control 47:510–519

Plainsirichai M, Turner DW (2010) Storage temperature during the early stage of ripening determines whether or not the peel of Banana (*Musa* spp. AAA, Cavendish subgroup) degreens, fruits. Cirad/edp Sci 65:69–74

Raju R (2008) Supply chain management, DBA 1730, IV semester course material. Centre of Distance Education, Anna University, Chennai, India

Saltveit ME (2004) Respiratory metabolism. Agriculture handbook number 66, U.S. Department of Agriculture

SARDI (2006) Maintaining the cold chain: air freight of perishables. South Australian Research and Development Institute

Shema A (2010) Shelf life studies: basics-concepts-principles. Mocon's advanced packaging solutions

Chapter 8
Food Traceability

8.1 Traceability Defined

Golan et al. (2004) mentioned that the definition of traceability is necessarily broad because traceability is a tool for achieving a number of different objectives and food is a complex product. Accordingly, several definitions of traceability and its classifications which come from organizations, legislations, and research literature can be found. According to ISO 8402 (1994) quality standards, traceability is defined as follows: "the ability to trace the history, application or location of an entity by means of recorded identification". In ISO 9000 (2005) standards, the definition is extended into "the ability to trace the history, application or location of that which is under consideration". ISO guidelines further specify that traceability may refer to the origin of materials and parts, the processing history, and the distribution and location of the product after delivery. Olsen and Borit (2013) define traceability as "the ability to access any or all information relating to that which is under consideration, throughout its entire life cycle, by means of recorded identifications". Obviously, traceability is a generic term which is applicable to products across industries; however, it is widely practiced in food industry.

The European Union (EU) regulation 178/2002 (EU 2002) narrows down the definition to food industry defining traceability as the ability to trace and follow a food, feed, food-producing animal or substance intended to be, or expected to be incorporated into a food or feed, through all stages of production, processing, and distribution. Codex Alimentarious Commission (CAC 2005) defines more concise definition of traceability as the ability to follow the movement of a food through specified stage(s) of production, processing, and distribution. ISO 22005 (2007) follows Codex definition of traceability and defines general principles and basic requirements for the design and implementation of a feed and food traceability system.

For agro-based food chain, Wilson and Clarke (1998) defined food traceability as the information necessary to describe the production history of a food crop and any subsequent transformations or processes that the crop might be subject to on its journey from the grower to the consumer's plate. Bosona and Gebresenbet

© Springer Nature Switzerland AG 2023
M. M. Aung and Y. S. Chang, *Cold Chain Management*, Springer Series in Advanced Manufacturing, https://doi.org/10.1007/978-3-031-09567-2_8

(2013) revised the definition as "a part of logistics management that capture, store, and transmit adequate information about a food, feed, food-producing animal or substance at all stages in the food supply chain so that the product can be checked for safety and quality control, traced upward, and tracked downward at any time". As there are various definitions of food traceability in research literature, Islam and Cullen (2021) undertook extensive literature review and revised the definition of food traceability as "an ability to access specific information about a food product that has been captured and integrated with the product's recorded identification throughout the supply chain".

Regarding the principles of traceability that might include characteristics, classification and components, an independent food safety watchdog, Food Standard Agency (FSA 2002) identified three basic characteristics for traceability systems: (i) identification of units/batches of all ingredients and products, (ii) information on when and where they are moved and transformed, and (iii) a system linking these data. Golan et al. (2004) suggested that an efficient traceability system should be characterized by breadth (i.e., the amount of information collected), depth (i.e., how far back or forward the system tracks the relevant information), and precision (i.e., degree of assurance to pinpoint a particular movement of a food product) to be able to make balance cost and benefits. According to Perez-Aloe et al. (2007), three are different types of traceability; namely back traceability or suppliers' traceability; internal traceability or process traceability; and forward traceability or client traceability depending on the targeted activity in the food chain. Typically, traceability has two components: tracking and tracing, where tracking is the ability to pinpoint the location and navigation of a particular product (in unit or a batch) following its path from the point of production to the final point of sale or point of consumption. Tracing allows to follow and verify the history of a product, its related activities and movement along the entire supply chain.

8.2   Regulations, Standards, and Guidelines for Food Quality and Safety

Food security refers to the availability of food and one's access to it. More specifically, access to sufficient, safe, and nutritious food that meets their food preferences and dietary needs is fundamental in sustaining life and promoting good health. Food safety, nutrition, and food security are closely linked, and unsafe food creates foodborne diseases and malnutrition, particularly affecting infants, young children, elderly, and the sick. Food safety assurance involves the reduction of risks which may occur in the food, while food quality assurance maintains the totality of features and characteristics of a product that bear on its ability to satisfy the standards or needs of the customer.

Due to globalization in food trade, ensuring food safety and quality is a matter of international significance and a responsibility of food producers and governments.

The global concern for food safety and quality; and the need for traceability are being addressed by the development of legislation, new international standards, and industry guidelines (Petersen 2004). To build customer confidence and to achieve safety and quality, participants in food supply rely on two methodologies. One manages food supply chains via regulations/standards or certifications. The second records logistics operations and production processes via a food traceability system that provides transparent traceback and track forward information (Hong et al. 2011).

The main purpose of food quality and safety legislation is to protect human health, to minimize environmental implications and to achieve fair competition by establishing uniform standards. Legislative systems can be classified into different levels (Trienekens 2004):

- worldwide level, e.g., Codex Alimentarius by CAC (FAO&WHO);
- regional level, e.g., EU's general food law;
- national level, e.g., (food legislations in the Netherlands, in India, etc.);
- industry-specific level, e.g., dairy products, seafood products, etc.

The following text highlights the development of worldwide and regional legislations, standards, and industry guidelines.

International organizations such as CAC established by the FAO, World Health Organization (WHO) and the International Standardization Organization (ISO) play an important role in the development of international standards and industry guidelines for food traceability (Petersen 2004). Over 180 states and the EU, as a member organization, work together in the general subject and commodity committees of which 17 are currently active. In its fifty-year existence, the CAC has adopted over 330 standards, guidelines, and codes of practice (BMEL 2021).

In 1993, The CAC recommended hazard analysis critical control point (HACCP) as the most effective system to maintain the assurance of a safe food supply (Beulens et al. 2005). Traditional food control procedures such as good hygiene practices (GHP) and good manufacturing practices (GMP) are accepted as prerequisites or the foundation for HACCP in an overall food safety management program (Huss et al. 2004).

In regional level, EU directive 178/2002, which enforces strict legislation on labeling systems for food products, US Bioterrorism Act of 2002, and FDA's Food Safety Modernization Act are leading legislative efforts to implement traceability of food. The important approach, "one-step-up/one-step-down" traceability enables actors in the food chain to identify the immediate supplier of a product as well as immediate subsequent recipient. This approach is the basic requirements for the design and implementation of a feed and food traceability system which is mentioned in EU regulation, ISO/DIS 22005, and the Bioterrorism Act 2002 of US (Ruiz-Garcia et al. 2010).

ISO is the world's largest developer and publisher of international standards. ISO has published over 24,000 standards covering various industries. The most widely known and impactful ISO standard is the ISO 9000 series for Quality Management Systems (QMS) in production environments which are independent of any specific industry. The 2000 version ISO 9001 (2000) addressed the standard model for quality management and quality assurance but did not address food safety. ISO 22000 (2005) specified requirements for a Food Safety Management System (FSMS) where an organization in the food chain needs to demonstrate its ability to control food safety hazards by enabling traceability. This standard requires to implement prerequisite programs and HACCP in combination with quality management system from ISO 9001 (FMRIC 2008). Furthermore, ISO 22005 (2007) defined the principles and objectives of traceability and specified the basic requirements for the design and implementation of a feed and food traceability system. It can be applied by an organization operating at any step in the feed and food chain. The strong feature of ISO 22000 standard is allowing successful system integration with other general management systems such as ISO 9001 and ISO 14001.

Another global standard is FSSC 22000 which is developed for the certification of food safety management systems for food manufactures and recognized by global food safety initiative (GSFI). In fact, this certification scheme is an integrated standard that focuses on sector-specific requirements and based on the requirements of ISO 22000 and ISO 9001. GSFI is the benchmarking body for the harmonization of international food safety standards, along with other food safety management schemes like the British Retail Consortium, International Food Standard, and Safe Quality Food schemes reducing trade barriers (https://www.fssc22000.com/scheme/). Regarding traceability, the GS1 system is particularly well suited to be used for traceability purposes and supports the global adoption of electronic product code information services which is a standard designed to enable electronic product code-related data sharing within and across enterprises (EPCglobal 2009).

There are other industry guidelines to follow in implementation of food safety. In the early 1990s, WHO developed the Ten Golden Rules for Safe Food Preparation, and more simpler version "Five Keys to Safer Food poster" was introduced in 2001.

The Five Keys to Safer Food are as follows:

(1) keep clean;
(2) separate raw and cooked;
(3) cook thoroughly;
(4) keep food at safe temperatures; and
(5) use safe water and raw materials (WHO 2006).

Temperature control is very important for food safety as microorganisms can multiply very quickly if food is stored at room temperature. The simple rule is to keep the high-risk food at temperatures below 5 °C or above 60 °C, where the growth of microorganisms is slowed down or stopped. One of the alternative methods of temperature control is referred to as the 2-h/4-h rule. The rule is to follow when the food is kept in temperature danger zone of 5–60 °C. The total time includes all the time the food has been at room temperature, for example during delivery, preparation, and transportation (SA Health 2009). The rule to follow is stated in Table 8.1.

Table 8.1 2-h/4-h rule

Total time (cumulative) between 5 and 60 °C	Action taken
Under 2 h	OK to use or refrigerate at 5 °C or less
2–4 h	OK to use straight away but cannot go back in the fridge
Over 4 h	Throw away

8.3   The Link Between Traceability and Quality and Safety

Consumer perceptions show an increasing concern about food safety and properties of the food they buy and eat. The information available from labeling conventions does not always translate into more confidence. It has been recognized that there is an increasing need for transparent information on the quality of the entire food chain, supported by modern tracking and tracing methods. Essentially, food quality is associated with a proactive policy and the creation of requirements to maintain safe food supply (Beulens et al. 2005). Product-tracing systems are essential for food safety and quality control. Traceability systems help firms isolate the source and extent of safety or quality control problems. The more precise the tracing system, the faster a producer can identify and resolve food safety or quality problems (Golan et al. 2004). In themselves, traceability systems neither produce safer/high-quality products nor determine liability. But they act as an element of any supply-management or quality/safety control system so that they can provide information about whether control points in the production or supply chain are operating correctly or not. So early detection and faster response to these problems are possible.

Quality and safety are both linked to traceability, whereas safety is implicated by traceability more often. They are two very important elements of people's conceptions of food and associated decision-making (i.e., food choice). Traceability is primarily viewed as a tool for the food safety by providing means for recall as well as proof for authenticity of food, but it also related to food quality. Since both quality and safety were shown to be related to confidence, traceability may indeed boost consumer confidence through quality and safety assessments (Rijswijk and Frewer 2006). Moe (1998) mentioned that traceability is an essential subsystem of quality management. Thus, a well-developed internal traceability system is necessary for quality management. It would efficiently improve data collection, production flow control, and quality assurance. It is a crucial system to fight against counterfeiting and improve operational efficiency by reducing resource wastage. GS1 (2021) stated that traceability systems should be supported by best practices, based on evolution of industry's needs, international regulations, and global standards. System complexity may vary depending on its placement along the supply chain (i.e., producer, manufacturer, distributor, retail, etc.), the product's characteristics, and the required business objectives.

FAO (2011) mentioned that quality and safety assurance are the responsibility of every actor in the supply chain and needed to share responsibility for implementing measures to prevent and control food contamination and deterioration. To ensure food quality and safety, quality and safety management systems play in very important role, and it should include the following:

- implementation of Good Practices (GAP, GMP, GHP etc.) throughout the production chain;
- application of HACCP system principles;
- implementation of management system that practices the principles of quality and safety management with continual improvement.

"Quality management principles" are a set of rules that can be used as a basis for quality management (ISO 2015). The seven quality management principles that can be applied are as follows:

- QMP 1—customer focus;
- QMP 2—leadership;
- QMP 3—engagement of people;
- QMP 4—process approach;
- QMP 5—improvement;
- QMP 6—evidence-based decision-making;
- QMP 7—relationship management.

ISO 9000, ISO 9001 and related ISO quality management standards are based on these seven QMPs. The adoption of the HACCP system is recommended by CAC to ensure food safety. Good Practices are the foundation to implement HACCP system which is an important component of FSMS.

To foster continuous improvement in the quality of products and processes, the firms use total quality management (TQM) system. The quality and safety management systems are the essential pillars to implement TQM. Ho (1994) stated that ISO 9000 can be seen as a route to implementing TQM. In ISO 9001:2015 standard, the requirement of identification and traceability is stated. An essential aspect of ISO management standards is continuous improvement. Organizations can use the Plan-Do-Check-Act (PDCA) cycle as described in ISO 9001 and ISO 22000 to drive this improvement. The PDCA model can apply to the overall FSMS and each of its elements. Figure 8.1 shows an integrated approach that comprise of safety, quality and traceability systems to implement TQM.

8.4 Food Contamination and Traceability

Foodborne disease outbreaks and incidents, including those arising from natural, accidental, and deliberate contamination of food, have been identified by the WHO as major global public health threats of the twenty-first century (WHO 2007). Many outbreaks are the consequence of a failed process, or inappropriate storage conditions

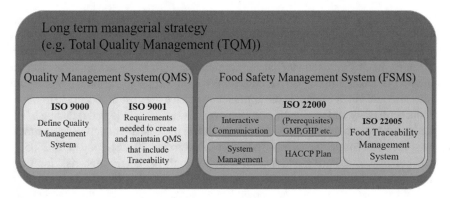

Fig. 8.1 Food safety, quality and traceability: an integrated approach

(usually temperature abuse) during distribution, food service or by the consumer. Most of these problems have been caused by the unintentional contamination of food, but there is a growing concern for the threat of intentional contamination such as bioterrorism. Food contaminants are substances that may be present in certain foodstuffs due to environmental contamination, cultivation practices, or production processes. Food may be accidentally or deliberately contaminated by microbiological, chemical, or physical hazards. In addition, there are other hazards/factors which cause contamination to food such as genetically modified organism and radioactive substances.

In the event of food outbreak and incidents, a traceback investigation is the method used to determine and document the distribution and production chain and the source(s) of a product that has been implicated in a foodborne illness investigation. Public health agencies conduct traceback activities to determine the source and distribution of the implicated product associated with the outbreak and to subsequently identify potential points where contamination could have occurred. This action helps prevent additional illnesses by providing a foundation for recalls of contaminated food remaining in the marketplace and identifying hazardous practices or violations. A traceback investigation may result in a recall of product (i.e., traceforward), other regulatory actions such as detention of an imported product, an injunction against a processor or grower, informing the public via press releases, closer monitoring of the product in general, domestic and foreign outreach, and "on-the-farm" investigations. Some of the challenges found in fresh produce tracebacks include the absence of labeling and distribution records, complex distribution systems, and multiple sources of product at the point of service. Another challenge is that traceback investigations are very resource intensive and may implicate but not confirm the cause of the contamination. These challenges include the fact that the epidemiology of foodborne disease is changing and new pathogens have emerged, some spreading worldwide (Guzewich and Salbury 2001).

WHO is promoting the use of all food technologies which may contribute to public health, such as pasteurization, food irradiation, and fermentation (WHO 2007). Also, the implementation of HACCP system is recommended to prevent food contamination by identifying potentially unsafe links in the food processing chain. The system manages the risk associated with food safety aspects of production (Kumar and Budin 2006). By having crisis management program that defines the action to be taken in the event of recall, impact can be reduced. For food companies, reducing processing batch size and batch mixing is the approach to reduce the cost of recalls, in terms of products quantity and media impact. However, it was also found that reducing batch size leads to losses in production efficiency, due to increased production setup times, setup costs, cleaning efforts, etc. (Saltini and Akkerman 2012; Depuy et al. 2005).

Especially, monitoring and surveillance for high-value and high-risk food are important, and inspection should be done at the port of entry, the best place to control food safety for imported foods. For preventative purposes, the analyses and interpretation of foodborne disease surveillance data require an associated and similar approach for data from food monitoring. The most modern and scientific way to perform is to use the risk assessment process that evaluates potential health risks to humans and animals. The integration of both foodborne disease surveillance and food monitoring could provide the data which are crucial for risk assessment (Schlundt 2002). Actually, traceability's strength lies in preventing incidence of food safety hazards and reducing the enormity and impact of such incidents by facilitating the identification of product(s) and/or batches affected, specifying what occurred, when and where it occurred in the supply chain, and identifying who is responsible (Opara 2003).

8.5   Logistic and Qualitative Traceability

The main fact that differentiates food supply chains from other chains is that there is a continuous change in the quality from the time the raw materials leave the grower to the time the product reaches the consumer (Apaiah et al. 2005). Perishables such as produce, meat, fish, milk, and more can change hands many times before reaching to the consumer. Keeping food in safe and good quality is a significant challenge as it moves through the supply chain. The quality of food is dependent on how food products are handled at every touch point throughout the food chain.

The efficiency of a traceability system depends on the ability to track and trace each individual product and distribution (logistics) unit, in a way that enables continuous monitoring from primary production (e.g., harvesting, catch, and retirement) until final disposal by the consumer. Traceability schemes can be distinguished in two types: logistics traceability which follows only the physical movement of the product and treats food as commodity and qualitative traceability that associates additional information relating to product quality and consumers safety, such as preharvest and postharvest techniques, and storage and distribution conditions (Folinas et al. 2006).

The food chain which demands for both logistics and qualitative traceability is found to be cold supply chain in which foods are perishable items and very sensitive to environmental conditions such as temperature, humidity, and light. The ability to collect this information and use it to ensure product quality in "real time" provides tangible benefits to the food industry. It provides a greater assurance of product quality and enables quick identification of problems which, therefore, can reduce food waste and spoilage. It also provides the mechanism for communicating to the consumer the diligence with which a business operates (Wilson and Clarke 1998). Papetti et al. (2012) presented a system called web-based infotracing system which integrates an electronic tracing system with a nondestructive quality analysis system. The web-based system was designed to acquire both logistics and qualitative information to the final consumer or to different food chain actors before or after purchasing, using the Radio Frequency Identification (RFID) code to identify the single and specific cheese product.

Transparency of a supply chain network is important as all the stakeholders of network have a shared understanding of access to product and process-related information they requested without loss, noise, delay, and distortion. Transparency enables to achieve efficient recalls on the chain level when necessary and support early warnings in case of a possible emerging problem through proactive quality monitoring system to optimize the supply chain (Beulens et al. 2005).

Based on the requirements of traceability in a food chain, a conceptual framework is considered (Fig. 8.2). In this framework, all supply chain actors are considered to have internal and external traceability to implement the whole supply chain traceability. The safety and quality regulations enforce actors to apply safety and quality assurance systems that comply with regulations and to manage all their operations

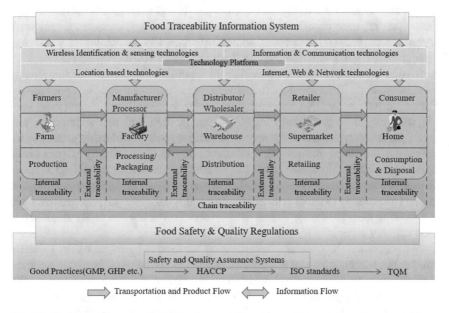

Fig. 8.2 Conceptual framework of a food traceability system (Adapted from Aung and Chang 2014)

in an efficient and standard manner. For supply chain operation and performance, enabling technologies can be seen as facilitators which serve as a medium for all actors to enable access to food traceability information system.

8.6   Technologies Applied

Opara (2003) mentioned the need of technologies for product identification, information capture, analysis, storage and transmission, as well as overall system integration. These technologies include hardware (such as measuring equipment, identification tags, and labels) and software (information systems). Advances in information and computer technology for information systems management; scanning and other digital technology for product identification, image capture, storage, and display; nondestructive testing and biosensors for quality and safety assessment; and geospatial technologies (geographic information system, global positioning system (GPS), remote sensing) for mobile assets tracking and site-specific operations are technological innovations that can be applied in a traceability system. Basically, a product traceability system requires the identification of all the physical entities (and locations) from which the product originates where it is processed, packaged, and stocked, and so this includes every agent in the supply chain (Regattieri et al. 2007).

For identification, alphanumeric code, bar code, QR code, and RFID are applicable technologies, and RFID is found as the most cutting-edge technology for supply chain integrity and traceability. But the problem is still the high cost of tags used in these systems, even though the prices have decreased significantly in recent years (Aarnisalo et al. 2007). Moreover, some limitations like readability of RFID tags through metal, glass, and liquid are difficult to achieve 100% (Petersen 2004).

Several technologies which complement identification for verification already exist, particularly in the livestock industry, for implementing traceable supply chains. Future innovations in DNA fingerprinting, nanotechnology for miniature machines, iris scanning, nose-print matching, facial recognition and retinal imaging, and their integration into plant and livestock industries have considerable potentials for improving the speed and precision of traceability in food industry (Opara 2003; Smith et al. 2008). Hologram labels can be another alternative technology that can be used to provide tamper protection, theft prevention, counterfeiting protection, and product traceability (Mlalila et al. 2016). Aarnisalo et al. (2007) mentioned that there is a growing need for the use of real-time sensors for quality and safety assurance in food industry especially for perishable food products.

In traceability, the traceback investigation for food is found necessary to verify counterfeit, authenticity, and provenance of food in the event of frauds or commercial disputes. In Europe, food legislation is particularly strict, and traceability systems, based on product labeling, have become mandatory in all European countries. In USA, the US Congress mandated country-of-origin labeling for many food crops/products as a requirement (Smith et al. 2008). However, the implementation of these systems does not ensure consumers against fraud. Paper documents can be counterfeit, so alternative methods for genetic traceability systems based on products identification are needed (Dalvit et al. 2007).

It is found that modern analytical techniques, in particular molecular biology techniques, can determine the plant or animal species present in a foodstuff. These techniques can be categorized into two types: the physicochemical techniques, which use either the variation of the radioactive isotope content of the product, spectroscopy, pyrolysis, or electronic nose, and the biological techniques which use the analysis of total bacterial flora or DNA chips. Using above techniques will help in differentiating milk produced on a mountain from that produced on the plains, of determining the origin of various cheeses or various wines, or of identifying the geographical origin of other foods like oysters, meats, fish, olive oils, teas, or fruit juices (Peres et al. 2007). Indication of origin may only become a signal of enhanced quality if the source-of-origin is associated with higher food safety or quality (Loureiro and Umberger 2006).

The study on the application of these techniques to improve food traceability can be seen in the integrated TRACE project (2005–2009) which is sponsored by European Commission through under sixth framework program and the recent coordinated research project of joint FAO/IAEA program. The TRACE project is to develop cost-effective analytical methods integrated within sector-specific and -generic traceability systems that will enable the determination and the objective verification of the origin of food. Mineral water, meat, honey, and cereals samples are analyzed in order to develop methods for the determination of food origin labeled with protected designation of origin or protected geographical indication. To verify the food origin, the applicability of using different methods such as trace elements and isotopes methods, rapid and profiling methods, molecular biology methods, and Chemometrics is studied. The project also addressed the issues of European consumer perceptions, attitudes, and expectations regarding food production systems and their ability-to-trace food products, together with, consumer attitudes to designated origin products, food authenticity, and food fraud (Rijswijk et al. 2006; TRACE 2009). Also, joint IAEA/FAO program proposed the implementation of nuclear techniques such as isotope ratio analysis along with multi-element analysis and other complementary methods, for the verification of food traceability systems and claims related to food origin, production, and authenticity (IAEA 2011).

The main problem found to all these techniques is the need for the construction of data banks which are very necessary for them. Therefore, TRACE project explored a mapping process that reduces the need for commodity-specific databases by finding the correlation between the tracers in food and the local environment (i.e., geology and groundwater). TRACE also exploits geological and climatic maps that are available and maintained annually. Under joint IAEA/FAO project, a database that enables to link with other databases is preferably hosted to facilitate its sustainability in the longer term for partners to use in provenance studies.

8.7   Emerging Technology for Traceability: Blockchain

Blockchain technology has recently gained significant attention and hype as a disruptive technology as it facilitates a transparent, open, immutable, distributed, secure, and auditable ledger shared by multiple nodes over the network without relying on a trusted third party. The network continuously grows in the form of blocks, as more transactions (information and data) are added. Individual transaction data files (blocks) are managed through specific software platforms that allow the data to be transmitted, processed, stored, and represented in human readable form (Kamilaris et al. 2019). As the blocks add on, the system develops and these blocks link and form a blockchain network or an ordered list of blocks using cryptography, which contain transactions, smart-contract creation, and invocations (Nakamoto 2008). The data in any given block cannot be altered retroactively without alteration of all subsequent blocks, which requires consensus of the network majority, thus it is very safe to business operations. The decentralized and transparent nature of a blockchain network enables the creation of a non-refutable record of data and real-time information sharing among various participants. These features contribute to its extensive applications in various domains, such as insurance, finance, supply chains, health care, and fraud detection.

Recent studies (Song et al. 2019; Mirabelli and Solina 2020; Kamble et al. 2020; Sunny et al. 2020) stated that blockchain is a promising technology to use in various supply chains: food/agricultural, pharmaceutical, express delivery, e-commerce, construction and automotive, etc., making supply chains sustainable, transparent, resilient, and efficient. Blockchain improves transparency, traceability, and auditability in supply chains tackling centralized system's challenges such as building collaboration and trustworthiness among actors, data manipulations, and cost entry barriers in legacy systems.

The new digital initiatives that adopt blockchain are started to simplify the processes and gain seamless integration in the entire supply chain. Maersk initiated a blockchain initiative titled global trade digitization intending to share information digitally and securely between maritime parties. Among many initiatives, TradeLens system led by Maersk and IBM seems to have formed an industrial standard, a step toward seamless integration throughout the transport supply chain, and a chance for maritime transport to expand on the feeder side toward the last-mile deliveries (Hvolby et al. 2021).

Participants in a blockchain network can track the journey of an asset from its origin to destination with the information appended on the blocks in blockchain. To capture this asset-related information (location, properties, etc.), blockchain needs to depend on existing methods (e.g., Internet of Things (IoT), RFID, GPS, etc.). Once the data are available in the blockchain, participants can trace it (Sunny et al. 2020).

Since cold chains distribute temperature-sensitive products, cold chain must be constantly monitored, controlled, and well documented in each step along the supply chains to prevent contamination of product and risk of consumers. Applying block chain, every single product in the cold chains can be assigned a cryptographically unique identifier at the start of the cold chain. This unique, strongly encrypted code can then be entered into the shared data ledger, and the history of all products can be accessed on the blockchain archive (Vandevelde 2018). Any issues with product can be detected and resolved within a matter of hours, thus response times can be minimized and overall product safety improved. It is considered that the blockchain will innovate cold chain management when integrated with proper IoT technologies such as RFID, bar codes, and sensors.

In summary, blockchain is a digital distributed ledger technology with unique combination of features such as decentralized structure, distributed notes and storage mechanism, consensus algorithm, smart contracting, and asymmetric encryption to ensure network security, transparency, and visibility (Kamilaris et al. 2019). It appears very promising to use in traceability that is vital for supply chains including cold chain; however, the technology is still in its early stage toward maturity and needs to address the barriers and challenges in different dimensions (Mirabelli and Solina 2020; Sunny et al. 2020).

8.8 Implication and Future Directions

Moe (1998) estimated that demand for information along the food chain will be increased and it will set the higher requirements for well-structured traceability systems. Therefore, traceability will be emerged as a new index of quality and basis for trade in future. Customer demand for real-time information about the products they buy and eat also will grow, and it will be one of the competitive advantages of food industry marketing.

The use of mobile phones accelerates the age of ubiquity. The ability to check food safety in the hands of the consumer has become a reality by tagging products with RFID or bar codes that can be read with a mobile phone. Smart phones today could be the future handheld device for traceability because of its portability, mobility, and accessibility to Internet and application software support. Consumers can scan the code in the store using mobile phone camera or embedded mobile RFID reader so they can find out the product history at their fingertips and make purchase for safe and quality foods. They can even offer feedback to the farmer.

In the near future, RFID and sensor-based systems will be widely used, not only for tracking the goods but also for monitoring the quality of the products and the supply chain itself. This will enable to detect the spoilage of food products and enhance the continuity of food supply chain. Biosensors will most probably be used for various uses such as detection of mycotoxins, bacteriocides, allergens, and contaminating microbes (Aarnisalo et al. 2007). The advanced techniques like gas chromatography and electronic noses (i.e., a machine which can detect and discriminate among complex odors using a sensor array) will be increasingly used in the field of food quality management (Peris and Escuder-Gilabert 2009).

The Internet promises to be an important tool for food traceability. Web-based traceability systems will enable traceability chains for products to personal computers and smart phones of consumers based on access control level of consumer identification system. This will deliver real-time information to consumers on the quality and safety status of products and also permit speedy recalls when quality and safety standards are breached. The larger trend in future is the convergence of smart phones with the IoT (i.e., Internet-connected real-world objects). Devices such as smart phones essentially become sensor and RFID readers, which allow consumers to interact with real-world objects in a much more detailed manner. IoT plays a vital role for monitoring, tracking, and tracing the cold chain products in the supply chain. IoT-enabled smart container, for example, is equipped with RFID, GPS, and other sensors. For quality traceability, the application of AI techniques such as fuzzy classification and artificial neural network is beneficial to evaluate the food quality timely along the supply chain and provides consumers with these evaluating information, to mainly enhance the consumer experience and help firms gain the trust of consumers (Wang et al. 2017).

To minimize the foodborne hazards and incidents, the sustainable agriculture which can produce good crop yields using natural methods to feed the soil and reduce pests (e.g., organic farming) should be maintained in order to balance economic, environmental, and quality of life benefits not only for farmers but also for consumers as well. As a consequence of incidents happened in livestock food industry (e.g., BSE, FMD, bird flu, and swine flu, etc.), monitoring and inspection of feeding diet and health of animals will become a mandatory task to do as human and animals share one health and the cost of impact on food industry and consumer confidence is intangible. Since the environmental concerns in food supply chain grow, to design and implement an eco-friendly supply chain would be a new challenge.

Most of the previous research are found to focus on traceability until retail point of the food chain, thereby missing to trace the consumer part of the food chain. In terms of food safety, consumer segment is also important; therefore, traceability should be extended to consumers. Blockchain is a promising technology toward a transparent supply chain of food, with many ongoing initiatives in various food products and food-related issues as it can be one of the potential technologies to provide reliable and secure traceability, but many barriers and challenges still exist, which hinder its wider popularity among farmers and systems (Kamilaris et al. 2019).

Traceability comes at a cost. But the costs of not having it or having inefficient systems in place may be severe both for governments, consumers, individual companies, and the whole food industry. It should be integrated into existing business systems, logistical processes, quality programs, and food safety programs such as HACCP as we mentioned in Figs. 8.1 and 8.2. In Chaps. 6 and 7, we presented quality control and assurance methods which are applicable in cold chain especially for a climacteric fruit.

Proper temperature management was proposed to reduce the loss of quality in cold chain, and measurement using sensors was recommended to access the level of quality in real time. However, the above scheme needs to keep quality history of the product as it can be used in turn by traceability system to detect food safety and quality problems so traceback could be done and consumers' confidence could be achieved. In conclusion, food traceability from "farm to fork" is going to become a reality if market forces, consumer demand, and government regulation all are converging to push a new level of supply chain visibility.

References

Aarnisalo K, Heiskanen S, Jaakkola K, Landor E, Raaska L (2007) Traceability of foods and foodborne hazards. VTT Technical Research Centre of Finland, Research Notes 2396

Apaiah R, Hendrix E, Meerdink G, Linnemann A (2005) Qualitative methodology for efficient food chain design. Trends Food Sci Technol 16(5):204–214

Aung MM, Chang YS (2014) Traceability in a food supply chain: safety and quality perspectives. Food Control 39:172–184

Beulens AJM, Broens DF, Folstar P, Hofstede GJ (2005) Food safety and transparency in food chains and networks. Food Control 16(6):481–486

BMEL (2021) Understanding food safety: facts and background. Federal Ministry of Food and Agriculture, Retrieved from https://www.bmel.de/SharedDocs/Downloads/EN/Publications/Und erstandingfoodsafety.pdf?__blob=publicationFile&v=4. Accessed 22 Feb 2022

Bosona T, Gebresenbet G (2013) Food traceability as an integral part of logistics management in food and agricultural supply chain. Food Control 33:32–48

CAC (2005) Codex procedural manual, 15th edn

Dalvit C, Marchi MD, Cassandro M (2007) Genetic traceability of livestock products: a review. Meat Sci 77:437–449

Depuy C, Botta-Genoulaz V, Guinet A (2005) Batch dispersion model to optimise traceability in food industry. J Food Eng 70:333–339

EPCglobal (2009) Retrieved from http://www.gs1.org/sites/default/files/docs/architecture/architect ure_1_3-framework-20090319.pdf. Accessed 02 Feb 22

EU (2002) Regulation (EC) No 178/2002 of the European parliament and of the council

FAO (2011) Cost-effective management tools for ensuring food quality and safety for small and medium agro-industrial enterprises. Retrieved from https://www.fao.org/3/i2385e/i2385e02.pdf. Accessed 20 Mar 2022

FMRIC (2008) Handbook for introduction of food traceability systems. Food Mark Res Inf Cent 2nd Ed., Tokyo, Japan

Folinas D, Manikas I, Manos B (2006) Traceability data management for food chains. Br Food J 108(8):622–633

FSA (2002) Traceability in the food chain a preliminary study. Food Standard Agency, UK. Retrieved from http://www.adiveter.com/ftp_public/articulo361.pdf. Accessed 15 May 2021

Golan E, Krissoff B, Kuchler F, Calvin L, Nelson K, Price G (2004) Traceability in the U.S. food supply: economic theory and industrial studies. Agricultural economic report number 830

GS1 (2021) GS1 global traceability compliance criteria standard, release 4.1, Retrieved from https://www.gs1.org/docs/traceability/GS1_Global_Traceability_Compliance_Criteria_For_ Food_Application_Standard.pdf. Accessed 22 Mar 2022

Guzewich JJ, Salbury PA (2001) FDA's role in traceback investigations for produce. Food Saf Mag, Target Group (Dec 2000/Jan 2001)

Ho SKM (1994) Is the ISO 9000 series for total quality management? Int J Q Reliab Manage 11(9):74–89

Hong AH, Dang JF, Tsai YH, Liu CS, Lee WT, Wang ML et al (2011) An RFID application in the food supply chain: a case study of convenience stores in Taiwan. J Food Eng 106:119–126

Huss HH, Ababouch L, Gram L (2004) Assessment and management of seafood safety and quality, 10–11 FAO fisheries technical paper 444

Hvolby H-H, Steger-Jensen K, Bech A, Vestergaard S, Svenssson C, Neagoe M (2021) Information exchange and block chains in short sea maritime supply chains. Procedia Comput Sci 181(2021):722–729

IAEA (2011) Information document from 55th International Atomic Energy Agency (IAEA) general conference 2011, 19–23 Sep 2011. Vienna

Islam S, Cullen JM (2021) Food traceability: a generic theoretical framework. Food Control 123:107848

ISO 8402 (1994) Retrieved from http://www.scribd.com/doc/40047151/ISO-8402-1994-ISO-Def initions. Accessed 15 May 2019

ISO 9001 (2000) Retrieved from https://www.iso.org/iso-9001-quality-management.html. Accessed 02 Feb 22

ISO 22000 (2005) Retrieved from https://www.iso.org/standard/35466.html. Accessed 02 Feb 22

ISO 22005 (2007) Retrieved from https://www.iso.org/standard/36297.html. Accessed 02 Feb 22

ISO (2015) Quality management principles. Retrieved from https://www.iso.org/iso/pub100080. pdf. Accessed 02 Feb 22

Kamble SS, Gunasekaran A, Sharma R (2020) Modeling the blockchain enabled traceability in agriculture supply chain. Int J Inf Manage 52:101967

Kamilaris A, Fonts A, Preanfea-Boldú (2019) The rise of blockchain technology in agriculture and food supply chains. Trends Food Sci Technol 91:640–652

Kumar S, Budin EM (2006) Prevention and management of product recalls in the processed food industry: a case study based on an exporter's perspective. Technovation 26:739–750

Loureiro ML, Umberger WJ (2006) A choice experiment model for beef: what US consumer responses tell us about relative preferences for food safety, country-of-origin labeling and traceability. Food Policy 32:496–514

Mlalila N, Kadam DM, Swai H, Hilonga A (2016) Transformation of food packaging from passive to innovative via nanotechnology: concepts and critiques. J Food Sci Technol 53(9):3395–3407

Mirabelli G, Solina V (2020) Blockchain and agricultural supply chains traceability: research trends and future challenges. Procedia Manuf 42:414–421

Moe T (1998) Perspectives on traceability in food manufacture. Trends Food Sci Technol 9(5):211–214

Nakamoto S (2008) Bitcoin: a peer-to-peer electronic cash system. Retrieved from https://bitcoin. org/bitcoin.pdf. Accessed 28 Dec 2021

Olsen P, Borit M (2013) How to define traceability. Trends Food Sci Technol 29(2):142–150

Opara UL (2003) Traceability in agriculture and food supply chain: a review of basic concepts, technological implications, and future prospects. Food Agric Environ 1(1):101–106

Papetti P, Costa C, Antonucci F, Figorilli S, Solaini S, Menesatti P (2012) A RFID web-based infotracing system for the artisanal Italian cheese quality traceability. Food Control 27:234–241

Peres B, Barlet N, Loiseau G, Montet D (2007) Review of the current methods of analytical traceability allowing determination of the origin of foodstuffs. Food Control 18:228–235

Peris M, Escuder-Gilabert L (2009) A 21st century technique for food control: electronic noses. Anal Chim Acta 638(1):1–15

Petersen A (2004) Status of food traceability in the European Union (EU) and United States of America (US), with special emphasis on seafood and fishery products., Masters' Assignment, Danish Technical University, Denmark

Regattieri A, Gamberi M, Manzini R (2007) Traceability of food products: general framework and experimental evidence. J Food Eng 81(2):347–356

Rijswijk WV, Frewer LJ (2006) How consumers link traceability to food quality and safety: an international investigation. In: 98th EAAE Seminar "marketing dynamics within the global trading system: new perspectives", Greece, pp 1–7

Ruiz-Garcia L, Steinberger G, Rothmund M (2010) A model and prototype implementation for tracking and tracing agricultural batch products along the food chain. Food Control 21(2):112–121

SA Health (2009) 2 hour/24 hour rule explained, food safety factsheet. Government of South Australia

Saltini R, Akkerman R (2012) Testing improvements in the chocolate traceability system: impact on product recalls and production efficiency. Food Control 23:221–226

Schlundt (2002) New directions in foodborne disease prevention. Int J Food Microbiol 78:3–17

Smith GC, Pendell DL, Tatum JD, Belk KE, Sofos JN (2008) Post-slaughter traceability. Meat Sci 80:66–74

Song JM, Sung JW, Park TH (2019) Applications of blockchain to improve supply chain traceability. Procedia Comput Sci 162(2019):119–122

Sunny J, Undralla N, Pillai VM (2020) Supply chain transparency through blockchain-based traceability: an overview with demonstration. Comput Ind Eng 150:106895

TRACE (2009) Tracing the origin of food. Retrieved from http://www.trace.eu.org/. Accessed 25 Apr 2020

Trienekens JH (2004) Book chapter in the emerging world of chains and networks, bridging theory and practice. Reed business information. ISBN 9789059019287, pp 253–267

Vandevelde P (2018) Heat-proofing cold chains with blockchain. Retrieved from https://www.supplychainbrain.com/blogs/1-think-tank/post/28629-heat-proofing-cold-chains-with-blockchain/. Accessed 21 Apr 2021

Wang J, Yue H, Zhou Z (2017) An improved traceability system for food quality assurance and evaluation based on fuzzy classification and neural network. Food Control 79:363–370

Wang Z, Wan T, Hu H, Gong J, Ren X, Xiao Q (2020) Blockchain-based framework for improving supply chain traceability and information sharing in precast construction. Autom Construct 111:103063. https://doi.org/10.1016/j.autcon.2019.103063

WHO (2006) Five key to safer food manual. Department of Food Safety, Zoonoses and Foodborne Diseases. Retrieved from https://www.who.int/foodsafety/publications/consumer/manual_keys.pdf. Accessed 2 Feb 2022

WHO (2007) The World Health Report 2007: a safer future. Geneva

Wilson TP, Clarke WR (1998) Insights from industry food safety and traceability in the agricultural supply chain: using the Internet to deliver traceability. Supply Chain Manag 3(3):127–133

Chapter 9
Design and Implementation of a Smart Refrigerator: A Case Study

From the early 2000s, the idea of connecting home appliances to the Internet had been popularized and was seen as the next big thing. Big electronics companies launched the Internet-based smart refrigerator, which features a touch screen which provides daily recommendation of food recipes and ability to connect to other devices such as speakers and smart televisions. However, there is no guarantee that the fridge will be getting updates after a few years, and it is too expensive for most people. Therefore, there is a market demand for smart refrigerators that own essential features in affordable price since the use of smart appliances is increasing today. This chapter presents about a case study on the design and implementation of smart refrigerator which could give smart features comparable to typical dull and cheap refrigerator.

9.1 Smart Refrigerator: A Smart Appliance for Smart Home

Refrigeration and freezing are widely used methods of preserving the quality of foodstuffs and thus protect consumer health and reduce food waste. They become important components of our daily lives and provide benefits in a variety of industries, including food, health, and the indoor environment (IIR 2007). Refrigeration is important as it can keep food cool and fresh after slowing down the activity of bacteria and extend shelf life. Nowadays, a refrigerator has become a common household item and one of the most important pieces of equipment in the kitchen for keeping food safe and fresh. With attached freezer compartment, refrigerator allows people to store some food longer and eat it at leisure and does bulk purchases to save money.

The International Institute of Refrigeration (IIR) estimates that the total number of refrigeration, air-conditioning and heat pump systems in operation worldwide is roughly 5 billion, including 2.6 billion air-conditioning units (stationary and mobile)

© Springer Nature Switzerland AG 2023

M. M. Aung and Y. S. Chang, *Cold Chain Management*, Springer Series in Advanced Manufacturing, https://doi.org/10.1007/978-3-031-09567-2_9

and 2 billion domestic refrigerators and freezers. Global annual sales of such equipment amount to roughly 500 billion USD. Over 15 million people are employed worldwide in the refrigeration sector which consumes about 20% of the overall electricity used worldwide. Based on the number of refrigerated appliances installed and their electricity consumption, the IIR estimates that domestic refrigerators and freezers consume almost 4% of global electricity (IIR 2019).

With the development of Internet of Things technology as well as improvement of people's living standards, smart home is no longer out of reach, and it is gradually becoming a reality. The smart refrigerator arises to match with fast-paced lifestyle and plays a critical role. With the variety of fresh food stored in the refrigerators, it has brought great convenience to people's daily life. The emergence of intelligent smart refrigerators allows people to fully appreciate the convenience of the refrigerator and gives full play to the use of intelligent refrigerators.

Smart refrigerator not only facilitates the convenience of people, but also reduces the waste by supplying the food timely. People can use a computer, mobile phone, or other intelligent terminal connected to the network to remotely control an intelligent refrigerator and upload real-time information to a cloud storage, where the status/information about food storage is updated on a regular basis by connecting the sensor, controller, and other equipment through a wireless network connection (Davenport et al. 2019). The application of intelligent refrigerators will give the busy people a very convenient life, so that those people can put more time and energy into the work and study, let the life can be in an orderly way (Qiao et al. 2017).

With the advance of technology, it is a well-known fact that the fast-paced development and modern living has resulted in a change of people lifestyle toward less exercise and an unhealthy diet. The fast-paced life has led to an alarming consumption of junk food, expired products, or vegetables at homes.

The technological development nowadays has enabled the use of smart appliances and machines almost everywhere. The refrigerator is considered one of the most important appliances in building a smart home. It is being used in almost every place for the purpose of storing foods, drinks, and medicines at cold temperatures and in a sealed place to avoid exposure. So, it is important to build a smart refrigerator that enables visibility, connectivity, and user control without missing desirable features.

9.2   Common Issues and Challenges with Typical Refrigerator

- **Food Waste**

With the internationalization of the food supply chain, food loss and waste can occur at each stage of the global food value chain, from agricultural production to final consumption. Although mismanagement may happen at one point of the chain, food waste mostly occurs at the end of the chain, in retail stores or households (Ndraha et al. 2018). Retail represents a considerable amount of waste in the food supply chain. At the end of the food supply chain, final consumption including commercial and household accounts for as much as 40% of total food losses. Especially, in the developed countries, food is mainly wasted in the final consumer stage of the supply chain (Martin-Rios et al. 2018).

Food waste is becoming a big problem in America (Cook Smarts, nd). The average American family throws away over $2000 worth of food every year. They suggest ways to reduce food wastes such as make a meal plan and grocery list; cook ingredients with shorter shelf life first; turn leftovers into new meals and proper utilization of the freezer. According to Davenport et al.'s survey (2019), Americans throw out a lot more food than they expect they will, food waste that is likely driven in part by ambiguous date labels on packages, a new study has found. Those survey participants expected to eat 97% of the meat in their refrigerators but really finished only about half. They thought they would eat 94% of their vegetables but consumed just 44%. They projected they would eat about 71% of the fruit and 84% of the dairy, but finished off just 40 and 42%, respectively. No one knows what 'use by' and 'best by' labels and those who check nutrition labels frequently waste less food.

There are some reasons behind the occurrence of food waste. The modern living and the fast-paced environment do not allow the user to keep a track of the food items inside the refrigerator. The traditional refrigerator is difficult to figure out which food is surplus, which food has been used up, and which food is going to expire, if not physically check the refrigerator. Finding an expired item in the refrigerator or required item that is no longer available in the fridge is the common problem that everyone may face. The item could be a food item, a drink, or medicine. Leaving the fridge door open is another important issue, which by repetition might lead to damaging the refrigerator. Moreover, liquid's leak is another problem that can be encountered in domestic refrigerator (Ali and Esmaeili 2017).

- **Poor Visibility and Temperature Management**

Another fact is that the temperature at which a refrigerator operates is critical for the safe storage of chilled food. However, the owners are unknown to the actual temperature their refrigerator operated at, and most users believe that the value is what they set at the temperature dial inside. There is also a practice of setting the temperature according to the weather such as setting the refrigerator to a lower temperature in the summer. There are other situations where user has less awareness of the recommended refrigerator temperature setting; lack of practice for periodic temperature measurement and the design of refrigerator in which real-time data are not displayed. Many published surveys stated that many refrigerators throughout the world are running at higher than recommended temperatures (James et al. 2008; Dumitraşcu et al. 2020; Göransson et al. 2018). Aung and Chang (2014) highlighted that the proper control and management of temperature are crucial as the quality of the food can be degraded or contamination can happen if operating temperature is incorrect, not recommended, and not optimal. This is especially true in keeping a mix of food items that have different functional properties and temperature requirement.

- **Food Safety**

Time–temperature profile of refrigerators is a good food safety indicator as it shows the real fluctuations of the recorded temperatures and gives a better understanding of the food safety risks related to consumer practices. Typically, the most refrigerators available in market have no record keeping system for time–temperature profile, inventory updates, and detailed information about the items stored inside (Dumitraşcu et al. 2020). From the study of Ak-Kandari et al. (2019) about food handlers, it is found that most of the customers were not aware of the importance of the essential time and temperature control required for preventing the microbial growth in foods.

By storing food properly, consumers are the final line of defense in preventing food spoilage and ensuring food safety. Refrigerators form an important link in the cold chain and represent a significant vector for domestic foodborne illnesses (Dumitraşcu et al. 2020).

9.3   Why We Need a Smart Refrigerator?

With the improvement of people's living standards and the accelerating pace of people's life, the refrigerator is playing an increasing important role in our daily life for keeping foods such as raw or cooked food, eggs, meats, fruits, vegetables, drinks, and some medicine. Since it has brought great convenience to people's life, more and more foods are kept inside the refrigerator to consume at a convenient time. However, it also brings some issues that continues to increase. The following are some issues commonly found in domestic/traditional refrigerators:

- There is no notification system about the foods that are going to expire, then the food will be expired if it is not consumed in time and make it a waste as most consumers feel like the food that is expired is not fit for consumption.
- Most refrigerators are dull, and the user has no knowledge about the detailed information with the stock on hand of the food inside unless it is checked manually; therefore, it is not easy to replenish the stock or to prepare for cooking and may lead to duplicate/unnecessary buying.
- Most of the refrigerators currently in the domestic market are manually controlled and did not achieve remote wireless control. Also, there is no software system available for control/interaction between refrigerator and the users.
- The refrigerator is very important to keep food fresh and safe as it relates to people's daily diet. However, no record is kept in conventional refrigerator for time–temperature profile, quantity and quality information of food items and user activities (e.g., door opening/cleaning), therefore difficult to do correction and maintenance in time for abnormal conditions. In addition, there is no real-time monitoring system about health or condition of the refrigerator such as energy consumption, refrigeration system, odors and leaks, storage capacity status, and air flow inside.

- Refrigerator is used not only for keeping ready to eat food but also ingredients for recipes. If the cook has no knowledge on the stock available, it is a daunting task for him/her to decide what recipe to prepare/cook. In situation where recipes recommended based on the ingredients are available, it will save time and make the cooking very straightforward to start.

Nevertheless, conventional refrigerators have been performing great task in preserving food items for a period of time, but there is need for more efficient ways of preserving and managing food items and seamless connectivity with the users (Osisanwo et al. 2015).

With the development of Internet of Things (IoT) technology as well as improvement of people's living standards, smart home is gradually becoming a reality. With the variety of fresh food stored in the refrigerators, it has brought great convenience to people's daily life. The emergence of smart and intelligent refrigerators allows people to fully appreciate the convenience of the refrigerator and give full play to the use of intelligent refrigerators by adding desirable and useful features (Qiao et al. 2017).

It is expected not only to have convenience to the people but also to easily manage items or resources kept in it, save unnecessary cost, reduce food wastage with timely consumption, plan an organized menu, as well as organized shopping list. Internet-connected smart refrigerators available today offer many features and are beneficial to the user, but they are still expensive and have challenges such as privacy and security concerns. Therefore, a direction to build a smart refrigerator with essential functionalities and less expensive price offerings is a necessary step to develop the market.

Recently, many refrigerator manufacturers are developing smart refrigerators. Typical features of smart refrigerator are as follows:

- WIFI capability to interface with other devices;
- recipes and cooking information displayed by built-in screen;
- storage information (inventory information, expiration information, etc.) displayed by built-in screen;
- notes and schedules for family members;
- a smart TV capability, etc.

It is found that transport between purchase and household refrigerator was the most sensitive link with respect to temperature abuse (Ndraha et al. 2018). Specific process-oriented and technology-based innovations were frequently identified as suitable strategies for reducing waste production and improving waste management. Technology may also help in reducing food waste by dealing with leftovers or by analyzing practical shelf life (Martin-Rios et al. 2018). In general, smart refrigerator with intelligent management system may help to reduce the food waste in food supply chains. It can also help with the users to reduce the unawareness of safe refrigeration practices.

9.4   Development of Smart Refrigerators

The food industry has boomed due to the development of refrigeration, yet for more than 100 years it has used the same technology. It is said that more than 45% of world food production would spoil if it were not for cold storage and distribution (Belman-Flores et al. 2015). Developing smart appliances has become a trend in the era of IoT and is critical to the realization of a smart home. Both industry and research have focused on developing smart refrigerators/fridges which bring comfort and smart lifestyle to the consumers.

- **Development in Industry**

The first refrigerator connected to the Internet called Quantified Fridge was in a wired 100-year-old house in the Netherlands by Alex van Es in 1998 with features of connecting Internet, having mailbox, doorbell and keeping and broadcasting records for door opening. In 1999, Electrolux Screenfridge, a connected refrigerator was designed with functions for remote access, multimedia control capabilities and to allow users to order groceries over the Internet (Osisanwo et al. 2015).

The first commercial Internet refrigerator, the Internet Digital DIOS, was introduced by LG in 2000. However, the products were unsuccessful because the consumers had seen them as an unnecessary and expensive (high cost) product, and the features available did not address to typical consumers' use cases.

In 2002, Whirlpool attempts to change the traditional fridge function of storing food items in a cool environment by adding more functionalities, i.e., enabling inventory control, adding recipes, integrating fridge with TV, radio, and computer like capabilities and even connection to the internet (Luo et al. 2009).

In 2009, CISRO Australia built a prototype of a smart fridge which can enable better nutrition and enhance health. The fridge is designed to have intelligent multimedia capabilities through touch screen interface, and it can manage items stored in it and advise to its users with cooking methods depending on what kind of food is stored. In addition, it has functions such as dietary control, nutrition monitoring, and eating habit analysis (Luo et al. 2009).

In global market today, smart refrigerators are built in different sizes and functionalities by many companies to sell in the market. Product differentiations are made by price, design, storage, connectivity, and smart features. The trend is seen going to make multi-functional and AI-powered appliance as a next smart thing. Many smart fridges also use popular voice assistants like Alexa and Google Assistant. The Samsung's Family Hub, LG's InstaView, and GE's Café brands are some examples of the popular smart refrigerators.

According to the forecast of Global Market Insights Inc. (https://www.gminsights.com/) in 2019, the market valuation of smart refrigerators will cross $7.5 billion and number of shipments is expected to hit 2 million units by 2026. Prominent companies in the smart refrigerator market are LG Corporation, Samsung Electronics, Haier Group, Whirlpool Corporation, Panasonic Corporation, Siemens AG, Bosch Group, and Midea Group.

- **Research Works**

In recent years, the research community paid more attention to making smart fridge using RFID and sensors that can also connect to Internet for easy remote control, monitoring, and information retrieval. Other than designing and making an expensive new fridge, most of the research recommend building the embedded system/module that can transform the standard refrigerator into a smart one. The smart refrigerator built has the functions of managing the food, controlling the temperature inside the refrigerator, browsing the recipe, and shopping online for replenishment (Qiao et al. 2017). Infrared (IR), proximity, pressure, odor, and door sensors are used to know weights, to detect low stock, leakages, and door status. Light-dependent resistors (LDR) sensor is used to monitor precisely the status of food items and notify the users about scare products via SMS and email, allowing to order by providing a link of the online vendor for particular items (Prapulla et al. 2015; Likitha et al. 2016; Sangole et al. 2017). Expired date and inventory information of items in the refrigerator are accessible by users and designed to alarm the user by email/SMS through mobile phone if any liquid leak happens or if the refrigerator's door is left open for a long time (Ali and Esmaeili 2017). Furthermore, voice-assisted instruction system can be developed for man–machine interaction through mobile application to enable to see the condition or receive notification about the food items kept inside the refrigerator. Thus, it will save the money and food wastage as well as help us to live a healthier lifestyle (Shruti et al. 2018). The data collected from the sensor system can be stored in local server or the cloud storage through IoT interface (Velasco et al. 2019). With increasing use of AI and machine learning in system, the smart refrigerator becomes content aware and focuses on personalization for each of its users (Kumar et al. 2019).

9.5   Common Concerns About Smart Refrigerators

With all the features and connectivity, many people have concerns about whether buying a smart refrigerator is a smart decision. The followings are some of common concerns when it comes to making the investment in a smart refrigerator.

- **Energy Efficiency**

It is apparent that the energy consumption of refrigerators and freezers has fallen dramatically over the years as many manufacturers focused on thermal and energy performance (Harrington, 2017; Belman-Flores et al. 2015). Many efforts were made in the development of the smart refrigerator, however none of which has been energy efficient or cost effective (Prapulla et al. 2015). With additional embedded hardware and software to offer more functions, considerable energy efficiency does not come to be a point for buying decision.

- **High Price**

Although the price gap between smart refrigerators and ordinary refrigerators has shrunk in recent years, there is still a notable premium for smart refrigerators. Most of the smart refrigerators are too expensive compared to ordinary refrigerators costing upwards of $3000. Multi-functioned smart refrigerators are not modularly built; therefore, the price tag is not flexible for buyers who want to use smart features, but they have budget constraint.

- **Technology**

The technology is somewhat complicated or complex for a typical household user who has limited understanding of how the smart refrigerator functions. Most regions are still struggling with inadequate Internet connectivity and restricted network connectivity, resulting in either low Internet speeds or poor support. The smart home environment, often known as the networked home, lacks sufficient security to secure data flow out of the house. Attackers may jeopardize the user's and the house's privacy. There is no specific operating system for controlling the smart system using a remote device (Prapulla et al. 2015).

9.6   Design and Development

In this section, the design and development of a smart refrigerator which is built by modifying an ordinary refrigerator are introduced. Unlike building a whole new refrigerator that would be expensive, an embedded module can transform ordinary refrigerator into a smart one which is cost saving and capable of sensing and monitoring to the refrigerator as well as communicating to the user.

The system was developed with the support of the KIAT's R&D project (Lab to Market 2016). Figure 9.1 shows the overall concept of the smart refrigerator developed.

As in the figure, the smart refrigerator consists of ordinary refrigerator, RFID, bar code, WIFI AP, and PAD. The operational system was based on Windows 10, refrigerator management software was developed by JAVA with Maria DB (https://mariadb.org/), and for communication Bluetooth 4.0 and Ethernet were used. Mobile monitoring system allows connectivity from an external system depending on the user setting.

Figure 9.2 shows the overall function of smart refrigerator system developed by Lab to Market Inc (2016). The owner has control over refrigerator and its inventory in this system. Monitoring can be done through dashboard function, and users' information can be managed easily. Menu recommendation is available based on inventory information, and shopping list will be prepared to the user.

Figures 9.3 and 9.4 show the exterior of smart refrigerator which has PAD, RFID, and bar code system. As in Fig. 9.3, a mounting fixture was designed to fix PAD or

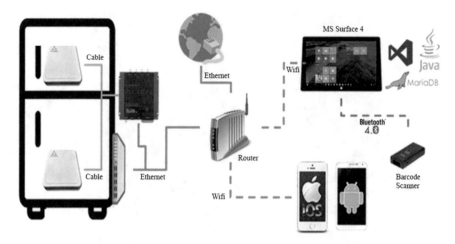

Fig. 9.1 Concept of smart refrigerator

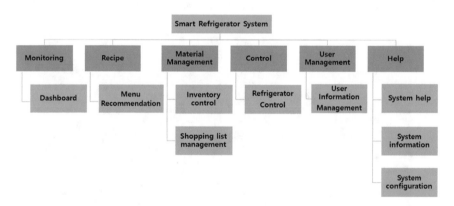

Fig. 9.2 Smart refrigerator system

mobile phone. PAD can be used with the refrigerator to display information about the refrigerator or to role as a PC whenever the user needs to use.

PAD can be fixed in the refrigerator and can be detached. Figure 9.4 shows top and rear view of the smart refrigerator, respectively.

As in the top view, RFID antenna is installed on the top of refrigerator in order to interrogate products for check-in (stored) and check-out (pick out). Once product ID is read by antenna, relevant information is collected from Product Database and displayed in the PAD. If product is not in the product database, one can register it using one of the three input methods: RFID, bar code, or manual input.

Figure 9.5 shows some components of refrigerator. As in the figure, the refrigerator consists of WIFI AP, controller, RFID reader, door sensor and power monitor, humidity sensor, etc.

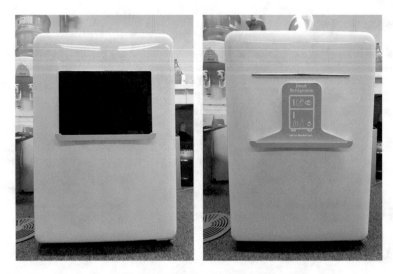

Fig. 9.3 Smart refrigerators with mobile PAD and without mobile PAD

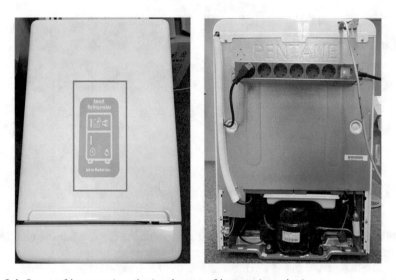

Fig. 9.4 Smart refrigerators (top view) and smart refrigerator (rear view)

Figure 9.6 shows the information architecture of smart refrigerator which consists of UI (i.e., smart refrigeration system UI), data storage, middleware, and smart refrigerator system:

- Data storage manages information generated by system (e.g., Log), food-related data (e.g., recipe, condition, etc.), and user information.

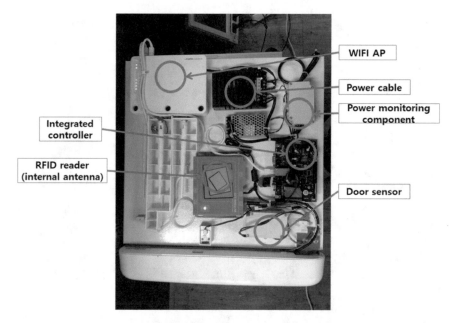

WIFI AP

Power cable

Power monitoring
component

Integrated
controller

RFID reader
(internal antenna)

Door sensor

Fig. 9.5 Components for smart refrigerator

- Automatic identification and data capture (AIDC) module collects information on the materials using various AIDC.
- Refrigerator manages cold storage and freeze storage.
- Middleware has two interface modules and controller for event handling: Equipment interface module (EQ interface module) controls temperature and supports interface with refrigerator, while AIDC interface module collects material information by interfacing with various AIDC.
- Smart refrigerator system consists of material manager, monitoring, and viewer: Material manager manages information on the materials kept by refrigerator; monitoring monitors real-time material location, analyzes information on specific material using data stored; viewer consists of user interface (UI) engine which provides system interface environment for user.
- Smart refrigerator system provides multi-platforms: Windows 10-based CS environment and smart phone-based mobile environment.

Conventional refrigeration system and recently developed smart refrigeration system store/manage all the materials with the same temperature even though different storage conditions are required for freshness. The smart refrigerator system in this section is designed to monitor and optimize different temperature requirements and the freshness of products. Figure 9.7 shows monitoring screen of environmental information such as temperature, humidity, and current. As shown in the left side of Fig. 9.7, the smart refrigerator has various functions such as dashboard, food list, food type, recipe, and recipe recommendation and also can store menu.

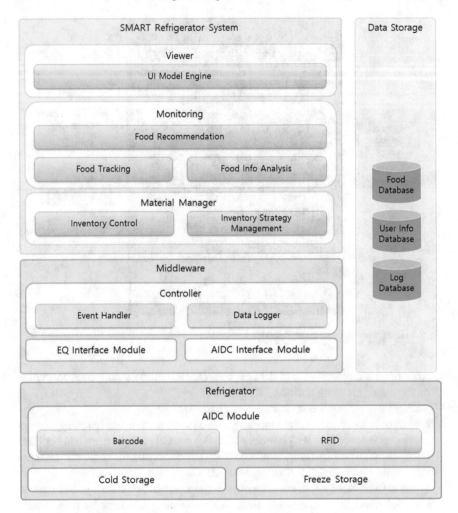

Fig. 9.6 Information architecture for a smart refrigerator

9.7   Summary

The world's first Internet refrigerator was an unsuccessful product because customers had seen it as an unnecessary, expensive, and luxury product (Aheleroff et al. 2020). Most researcher highlighted that one affordable solution for consumers is to upgrade/transform an existing refrigerator into a smart and cost-saving one. Other than building the refrigerator in fancy design with entertainment functionalities, adding only essential features will help to save cost and energy.

The ultimate objective of building the smart refrigerator is to make our daily life with food/kitchen more convenient. The communication between people and machines would make life easier with the advance of technology and wider use

Fig. 9.7 Monitoring screen: temperature, humidity, and power

of Internet. Remote control and monitoring of any system are desirable, and it is expected to manage items or resources kept in it, save unnecessary cost, save food wastage, plan an organized menu, as well as organized shopping list.

In summary, the design and development of smart refrigerator mentioned in this study are simple and modular. And the commercial grade components are used in the system unlike other prototypes, therefore, it is found robust and operational. Essential features are considered and designed for daily usability with easily manageable UI for monitoring and control. As a feature of smart refrigerator, power consumption status indicator is included, and the efficient record keeping and information system is necessarily implemented to enable traceability for unexpected food incidents. Having local database access enables user to get secure and fast access to device, and the system can run smoothly using local wireless network without Internet access. The system can be controlled wirelessly, and shopping list and menu recommendation are outstanding features that help the user to save time and effort.

It will take some years for newly build smart refrigerator to be affordable for many households such as smart TVs today. While they started out quite a bit more expensive, the prices have come down significantly as more brands and models have become available. Choosing a smart fridge over (non-smart) ones with a bottom-drawer freezer or a French-door style could cost as little a couple hundred bucks more or as much as a couple of thousand dollars more. It all depends on the model and brand that the user chooses.

References

Aheleroff S, Xu X, Lu Y, Aristizabal M, Velásquez JP, Joa B, Valencia Y (2020) IoT-enabled smart appliances under industry 4.0: a case study. Adv Eng Inform 43:101043

Ali Z, Esmaeili SE (2017) The design of a smart refrigerator prototype. In: 4th Int'l conference on electrical engineering, computer science and informatics (EECSI), 19–21 Sept 2017, Yogyakarta, Indonesia

Ak-Kandari D, Al-abdeen J, Sidhu H (2019) Food safety knowledge, attitudes and practices of food handlers in restaurants in Kuwait. Food Control 103(2019):103–110

Aung MM, Chang YS (2014) Temperature management for quality assurance of perishable food supply chain. Food Control, Elsevier, vol 40, pp 198–207

Belman-Flores JM, Barroso-Maldonado JM, Rodríguez-Muñoz AP, Camacho-Vázquez G (2015) Enhancements in domestic refrigeration, approaching a sustainable refrigerator—a review. Renew Sustain Energy Rev 51:955–968

Cook Smarts (nd) https://www.cooksmarts.com/cooking-guides/cook-on-a-budget/reduce-food-waste/. Accessed 10 July 2020

Davenport ML, Qi D, Roe BE (2019) Food-related routines, product characteristics, and household food waste in the United States: a refrigerator-based pilot study, 50, November 2019, Article 104440

Dumitraşcu L, Nicolau AN, Neagu C, Didier P, Maître I, Nguyen-The C, Skuland SE, Møretrø T, Langsrud S, Truninger M, Teixeira P, Ferreira V, Martens L, Borda D (2020) Time-temperature profiles and Listeria mnocytogenes presence in refrigerators from households with venerable consumers. Food Control 111:107078

Göransson M, Nilsson F, Jevinger Å (2018) Temperature performance and food shelf-life accuracy in cold supply chains- insights from multiple field studies. Food Control 86:332–341

Harrington L (2017) Quantifying energy savings from replacement of old refrigerators. Energy Procedia 49–56

IIR (2007) Refrigeration drives sustainable development: state of the art-report card. International Institute of Refrigeration

IIR (2019) The role of refrigeration in the global economy. In: 38th Informatory note refrigeration technologies. International Institute of Refrigeration

James SJ, Evans J, James C (2008) A review of the performance of domestic refrigerators. J Food Eng 87:2–10

Kumar JNA, Suresh S, Vaani KN (2019) An AI driven approach for smart refrigerator to enhance family diet and sustainability. In: Alliance international conference on artificial intelligence and machine learning (AICAAM), April 2019

Lab to Market (2016) Development of smart refrigerator considering product information, R&D re-discovering project, Unpublished Report. Project no: N053400067, KIAT, Republic of Korea

Likitha RV, Nagashree R, Shruthi P (2016) IoT smart fridge. Int'l J Adv Res Electron Commun Eng (IJARECE) 5(4):1136–1139

Luo S, Jin JS, Li J (2009) A smart fridge with an ability to enhance health and enable better nutrition. Int'l J Multimed Ubiquitous Eng 4(2):69–79

Martin-Rios C, Demen-Meier C, Gössling S, Cornuz C (2018) Food waste management innovations in the food service industry. Waste Manage 79:196–206

Ndraha N, Hsiao H-I, Valajic J, Yang M-F, Lin HTV (2018) Time-temperature abuse in the food cold chain: review of issues, challenges, and recommendations. Food Control 89:12–21

Osisanwo F, Kuyoro S, Awodele O (2015) Internet refrigerator—a typical Internet of Things (IoT). In: 3rd Int'l conference on advances in engineering sciences and applied mathematics (ICAESAM), Mar 23–24, London

Prapulla SB, Shobha G, Thanuja TC (2015) Smart refrigerator using internet of things. J Multidiscip Eng Sci Technol (JMEST) 2(7):1795–1801

Qiao S, Zhu H, Zheng L, Ding J (2017) Intelligent refrigerator based on Internet of Things. In: IEEE Int'l conference on computational science and engineering (CSE) and embedded and ubiquitous computing (EUC), 21–24 July 2017, Guangzhou, China, 406–409

Sangole KM, Nasikkar BS, Kulkarni DV, Kakuste GK (2017) Smart refrigerator using Internet of Things (IoT). Int J Adv Res Ideas Innov Technol 3(1)

Shruti L, Seethal S, Urmil S, Tejawi M (2018) A food management system based on IoT for smart refrigerator. Int J Res Appl Sci Eng Technol (IJRASET) 6(5):2304–2307

Velasco J, Alberto L, Ambatali HD, Canilang M, Daria V, Liwanag JB, Madrigal GA (2019) Internet of things-based (IoT) inventory monitoring refrigerator using Arduino sensor network. Indonesian J Electric Eng Comput Sci 18(1):508–515

Q&A of Chapters

Chapter 1: Introduction

1. Perishable goods are most vulnerable to _____ in nature.

   a) pressure
   b) light
   c) temperature
   d) scratches

2. The principal aim of CCM is _____.

   a) to assure product quality
   b) to obtain product safety
   c) to minimize wastage
   d) all of the above

3. The biggest challenge of temperature control in cold chain is

   a) expensive cold chain equipment
   b) diverse characteristics of perishables
   c) the lack of temperature measurement equipment
   d) packaging of perishables
   e) all of the above

4. A breakdown in temperature control cause the food

   a) decrease esthetic appeal
   b) reduce nutritional value
   c) spoil
   d) loss quality
   e) all of the above

© Springer Nature Switzerland AG 2023
M. M. Aung and Y. S. Chang, *Cold Chain Management*, Springer Series in Advanced Manufacturing, https://doi.org/10.1007/978-3-031-09567-2

5. The degradation of food due to temperature breakdown is easily visible for every food product.

   a) True
   b) False

6. Who are the stakeholders in cold chain business?

Chapter 2: Fundamentals of Cold Chain Management

1. Cold chain is commonly found in the following commodities except

   a) semiconductors
   b) food
   c) canned fruit
   d) pharmaceuticals

2. Successful cold chains require _____ at every step of the journey.

   a) effective planning
   b) quick communication
   c) the proper facilities
   d) the right technology
   e) all of the above

3. The three main features of cold chain logistics are the following except _____.

   a) product specific
   b) complexity
   c) high cost
   d) coordination

4. 3Q in the '3Q' principle are the following except

   a) query
   b) quantity
   c) quality
   d) quick

5. What are the features of cold chain logistics?
6. What are four parts of cold chain logistics?
7. What is '3T' principle in Cold Chain Logistics?
8. What are the two main differences between supply chain and cold chain?
9. What are the most favorable situations to use passive thermal packaging systems instead of active one?

Chapter 3: The Development of Cold Chain

1. What is the best technology to date to preserve and maintain food in freshness with no associated risks?

   a) pickling
   b) smoking
   c) drying
   d) refrigeration

2. _____ is an assurance that food will not cause harm to the consumer when it is prepared and/or eaten according to its intended use.

   a) food quality
   b) food traceability
   c) food safety
   d) food labeling

3. To ensure food safety in cold chain is a responsibility of _____.

   a) processors
   b) distributors
   c) customers
   d) all stakeholders

4. External food quality attributes include all of the following except

   a) appearance
   b) feel
   c) texture
   d) defects

5. The technologies that improve the visibility and delivery of food include all of the following except

   a) GPS
   b) RFID
   c) WSN
   d) ozonolysis

6. What conditions vaccines are sensitive to?
7. What types of vaccine need ultracold storage condition?
8. What is functionality of VVMs?
9. List the technologies that are beneficial to cold chain management.

Chapter 4: Cold Chain Management Essentials

1. Fresh fruits and vegetables show _____ metabolism.

   a) digestive
   b) respiratory
   c) vaporization
   d) excretory

2. _____ is an ethylene producer.

   a) Watermelon
   b) Peas
   c) Avocado
   d) Carrots

3. Apple is a _____ fruit to freezing injury.

   a) not susceptible
   b) most susceptible
   c) moderately susceptible
   d) least susceptible

4. Respiration rate in broccoli is

   a) low
   b) moderate
   c) very high
   d) extremely high

5. The required refrigerated temperature for pharmaceuticals is

   a) 0 to +5 °C
   b) +2.0 to + 8.0 °C
   c) +8 to +10 °C
   d) −5.0 to +2 °C

6. Mention five different preservation techniques for food.
7. What facilities are used to control in cold chain?
8. What are the three types of heat concerns to refrigeration?
9. What is the reason of precooling for goods?
10. List the names of refrigeration methods.
11. What are the steps in freezing process?
12. What are the differences between slow freezing and quick freezing methods?
13. What product characteristics need to be considered in cold chain?
14. What are the design consideration for a reefer unit?
15. What are the advantages and disadvantages of multi-temperature vehicle/trailer?
16. Why OSH is important for workers in refrigerated facilities?

Chapter 5: Cold Chain Monitoring Tools

1. Time–temperature indicator is based on _____ reactions except

    a) physical
    b) chemical
    c) logical
    d) microbiological

2. In RFID application, 13.56 MHz frequency is used for _____ distance.

    a) 10 cm or less
    b) 10 cm–1 m
    c) 3–6 m
    d) 100 m

3. What ISO/IEC standard is used for freight containers?
4. Which technologies can be used for cold chain monitoring?

Chapter 6: Temperature Management in Cold Chain

1. The concerns in mixed load or storage of fruits and vegetables are the following except

    a) temperature compatibility
    b) RH
    c) size
    d) ethylene sensitivity

2. Multi-temperature vehicles usually have _____ compartments that have separately controlled temperature.

    a) 2
    b) 3
    c) 4
    d) 5

3. Why we need optimal target temperature in cold chain?
4. What are applicable methods for getting optimal temperature in cold chain?

Chapter 7: Quality Assessment in Cold Chain

1. How many patterns are in the respiration of fruits?

    a) 2
    b) 3
    c) 4
    d) 5

2. For the breach of temperature threshold value, the following dimensions are commonly considered except

   a) how many?
   b) how high?
   c) how often?
   d) how long?

3. State quality assessment methodologies.

Chapter 8: Food Traceability

1. Effective traceability system should have the following characteristics except

   a) breadth
   b) accuracy
   c) depth
   d) precision

2. Which statement is incorrect about traceability?

   a) traceability is a tool for recall.
   b) traceability boost consumer confidence.
   c) traceability is an essential subsystem of quality management.
   d) traceability system produces safer/high-quality products.

3. Mention two types of traceability systems.
4. What are the opportunities and challenges of traceability systems?
5. What technologies could be used to build a traceability system?
6. What is 2 h/4 h rule and when it will be used?
7. Describe the unique characteristics of blockchain.

Chapter 9: Design and Implementation of a Smart Refrigerator: A Case Study

1. What are the issues with typical home refrigerators?
2. What are the desirable features with smart refrigerators?
3. What concerns are there using with smart refrigerators?

Answer Keys

Chapter 1

1. Ans: c
2. Ans: d
3. Ans: b
4. Ans: e
5. Ans: b

Chapter 2

1. Ans: c
2. Ans: e
3. Ans: a
4. Ans: a

Chapter 3

1. Ans: d
2. Ans: c
3. Ans: d
4. Ans: c
5. Ans: d

Chapter 4

1. Ans: b
2. Ans: c
3. Ans: c
4. Ans: d
5. Ans: b

Chapter 5

1. Ans: c
2. Ans: b

Chapter 6

1. Ans: c
2. Ans: b

Chapter 7

1. Ans: a
2. Ans: c

Chapter 8

1. Ans: b
2. Ans: d

Index

© Springer Nature Switzerland AG 2023   161
M. M. Aung and Y. S. Chang, *Cold Chain Management*, Springer Series in Advanced
Manufacturing, https://doi.org/10.1007/978-3-031-09567-2

Printed in the United States
by Baker & Taylor Publisher Services